Albert Jack is an English writer and historian who became something of a publishing phenomenon in 2004 when his first book *Red Herrings and White Elephants*, which explored the origins of well-known phrases in the English language, became a huge international bestseller. His other successful titles include *Shaggy Dogs and Black Sheep, Phantom Hitchhikers, Loch Ness Monsters and Raining Frogs, Pop Goes the Weasel* and *What Caesar Did for My Salad.*

THEY LAUGHED
AT GALILEO

How the Great Inventors
Proved Their Critics Wrong

Albert Jack

A Herman Graf Book
Skyhorse Publishing

First published in 2015 in the UK by Constable,
an imprint of Constable & Robinson Ltd

This edition first published in 2015 by Skyhorse Publishing

Skyhorse Publishing books may be purchased in bulk at special discounts
for sales promotion, corporate gifts, fund-raising, or educational purposes.
Special editions can also be created to specifications. For details, contact
the Special Sales Department, Skyhorse Publishing, 307 West 36th Street,
11th Floor, New York, NY 10018 or info@skyhorsepublishing.com.

Skyhorse® and Skyhorse Publishing® are registered trademarks of
Skyhorse Publishing, Inc.®, a Delaware corporation.

Visit our website at www.skyhorsepublishing.com.

10 9 8 7 6 5 4 3 2 1

Library of Congress Cataloging-in-Publication Data is available on file.

Cover design by Rain Saukas
Cover illustration credit by Thinkstock

Print ISBN: 978-1-62914-758-1
Ebook ISBN: 978-1-63220-236-9

Printed and bound by CPI Group (UK) Ltd, Croydon, CR0 4YY

This book is for Colin Willmott in Woking, England. He taught me to understand that anything is possible and never to believe you are not good enough to make a difference. I think that is what this book is all about (and perhaps my career).

Thanks Dad

ACKNOWLEDGEMENTS

I would like to thank Robert Smith and Hugh Barker for helping to bring this project together and the following team at Constable & Robinson for actually making it happen: Rod Green, Dominic Wakeford, Howard Watson and Clive Hebard.

Last, but who should be first, Geodey Weisner who, in conversation, unwittingly came up with the idea in the first place. If you don't like the book then it is his fault.

CONTENTS

INTRODUCTION

If the world should blow itself up, the last audible voice
would be that of an expert saying, 'It can't be done.'
                                              Peter Ustinov

Curiosity will eventually lead to innovation. Fortunately
we are an imaginative species that does a lot of wonder-
ing. Way back to when man first learned to walk upright
and began communicating with each other, by pointing
and shouting, we can find the earliest examples of this.
Somebody once thought, 'I know, we can move that heavy
rock, or dead buffalo, by rolling it along on tree trunks
because it is easier than dragging it over the ground.'
This, of course, led to the wheel. It must have been
around that time that some other clever soul worked out
that if he held some meat over that hot firey thing then
it tasted better. It seems basic but it was innovation.
Somebody somewhere decided to take the risk of burning their food down into ashes, as they knew the burning
logs did, just to see if it tasted any better. But I bet there
was someone else laughing at him and saying, 'Don't do
that, it's a terrible idea' (or whatever is was they would
have said back then.) And that's innovation too. That's
discovery and invention.

We have been doing it ever since in one form or
another and we have come a long way as a species

thanks to people who take risks and ignore the advice of wiser ones. And that, in a nutshell, is what this book is all about. You see, for all of our innovations and invention over the last 6,000 years, it is incredible to understand that the one thing that has not changed at all is the human brain.

Believe it or not, the prehistoric human brain was perfectly capable of understanding how to use Windows 8.1 and could easily have landed a rocket on the moon if only the information it was given was better evolved at the time. The brain itself was already fine and all it needed was programming. That, of course, is what has happened to it over the many years since. Man has programmed its brain to learn new and better ways of doing things. And curiosity has led it to evolve from pointing and shouting, fire and tree trunks into where we are now. It is curiosity that has led to invention and migration. 'I wonder what is over that hill over there? There may be water, possibly better vegetation. Maybe there are more of those rabbit things we like to eat? Let's go and have a look.' This would have taken them from caves and into man-made huts and so on and so on. And all the time, at every step of the way, somebody would also have been saying to them, 'No, no. That's a terrible idea. It will never work.' Or a mother shouted, 'Don't climb onto the back of that thing, Jonny, it's not safe. You will hurt yourself,' which was followed by WHAM! and 'I told you so.' But, as we all know, Jonny must have got right back on that horse.

More recently, in 1916, somebody said of the radio, 'The wireless music box is of no imaginable commercial value. Who would pay for a message sent to nobody in particular?' Well, that would have been a fair question back then but imagine a world without the radio. And the same was said of the television when it was dismissed as a novelty. 'American families will not sit around staring at a plywood box for hours at a time.' How wrong can you be? King Gillette thought that men would use a razor blade once or twice and then throw it away to buy a new one. His friends, who were all using cut-throat razors handed down from generation to generation, told him he was mad. And nobody took George Devol seriously when he invented the robotic arm and the entire industrial industry simply could not understand how to replace a man, or woman, standing at a bench with a spanner. Well, millions of men and women actually.

The telephone was dismissed as a meaningless toy and the Chief Engineer of the British Post Office actually said, 'We have perfectly good messenger boys, thank you.' The Chairman of IBM thought there would be a world market for only five computers. Luckily for them (and us) his son, and successor, had other ideas; and the jet engine, which has changed the lives of everybody, almost cost Frank Whittle his, but he didn't give up. The Beatles were told that guitar bands were on their way out and Elvis was dismissed as a truck driver. Firemen were advised to grow whiskers, make sure they were wet and then stuff them in their mouths before running into smoke-filled buildings. That was until 1916 when somebody finally agreed that Garrett Morgan's Safety Hood was a good idea after all. It had only taken him four years to convince the authorities.

And that is what this book is all about. It tells the stories of countless inventive and curious minds, and how somebody somewhere thought, 'Now, there must be a better way of doing things than this.' And then they went off and spent years, in some cases, working out how. And there were some accidents along the way too. A melted chocolate bar was responsible for the microwave oven and a lab accident led to safety glass. J. K. Rowling and Vladimir Nabokov were both told nobody would read their books, and Marilyn Monroe was advised to improve her typing skills.

Some sacrificed their lives for their invention. In fact, in the case of parachutes, thousands of them did. Marie Curie famously spent a lifetime experimenting with cures for cancer and died of cancer as a result, and Wan Hu was incinerated when he tried, for the first time, to reach for the stars. The man who invented the modern newspaper press died when he became trapped in one and the list of personal sacrifices, so that we can live in the modern way we do, is a long one. And it has been going on for a very long time. It's the only way humans would have discovered which berries were poisonous and which they could safely eat; what killed you when it was raw but kept you alive after you cooked it; and, of course, how cows produced milk that was safe to drink. And, for that matter, what did they actually think they were doing to the cow when they found that out?

To some intriguing questions there can be no answer but for countless others we know exactly who discovered what and how. So sit back and join me on a journey through the history of invention and innovation, and discover for yourself just what was going through the minds of these people and who knew a good idea when they saw one. And also discover who told them it would never work. After all, when he first suggested that the earth was not at the centre of everything, they laughed at Galileo.

Albert Jack
Bangkok

SCIENCE AND TECHNOLOGY

THE RADIO

It was during the summer of 1894 when an unknown twenty-year-old Italian by the name of Guglielmo Marconi called his parents into a room to show them how he could make a bell, on a far wall, ring by simply pressing a button. He had done so by using electromagnetic radiation, first introduced by the German physicist Heinrich Hertz in 1888. Once Marconi's father, a wealthy landowner, had checked for trickery (there were no wires) he handed over the contents of his wallet, enabling his son to buy the equipment he needed for some even more ambitious experiments.

Within a year Marconi was able to send and receive electronic signals over a distance of 1.5 miles, both around hills and through buildings. Convinced of the value of his invention, particularly to the military and the telegraph companies who were busy stringing wiring all over the world, Marconi wrote to Pietro Lacava, the Italian politician who had become the Minister for Post and Telegraphs in 1889, outlining his 'wireless telegraph' and requesting funding. Marconi never received a reply, although the document did turn up much later at the ministry with the words 'to the Lungara' scrawled across the top; a reference to the infamous lunatic asylum on Via della Lungara in Rome.

Meanwhile, the young Italian continued with his experiments, achieving ever-improving results over longer distances, and decided to travel to England in 1896 where he presented his ideas to William Preece, the Chief Electrical Engineer of the British Post Office, who had himself been experimenting with wireless transmission since 1892. Preece immediately recognized the value of Marconi's new technology and introduced it to the Royal Society during a lecture called 'Signalling through Space without Wires', which was given in London on 4 June 1897, the very same year that the esteemed President of the Royal Society, Lord Kelvin, had piously announced, 'Radio technology has no future.'

However, by early 1899 Marconi was transmitting wireless messages between Cornwall and France and in November of that year he was invited to America to demonstrate his equipment. On the return journey aboard the SS *St Paul*, Marconi and his assistants set up a transmitter and the passenger liner became the first in history to report its estimated arrival time, some 66 miles from the English coast. Having built a station at South Wellfleet in Massachusetts, on 18 January 1903 he famously connected the American President, Theodore Roosevelt, with the English King, Edward VII, in what was the first ever transatlantic wireless communication, using Morse code, between America and the United Kingdom.

Within a decade Marconi's company had built powerful transmitters on both sides of the Atlantic and was responsible for nearly all of the communication between ships and land, even establishing a nightly news service for sea captains to relay to their passengers. It was a Marconi wireless telegram that alerted the British police to the likelihood that the notorious murderer Dr Crippen was heading for Quebec aboard the Canadian Pacific liner SS *Montrose*, allowing detectives to board a faster ship and arrest him on his arrival on 31 July 1910. It was the first time wireless communication had ever been used to catch a killer. Marconi's wireless telegram station also received news of the sinking of the *Titanic* in April 1912, allowing messages to be relayed to other ships in the area and saving countless lives in the process.

> A rocket will never be able to leave the Earth's atmosphere.
>
> *New York Times*, 1936

As hard as it is to imagine now, it is quite possible that without Marconi's technology all lives would have been lost and the sinking of the *Titanic* may, today, remain a mystery, as nobody would ever have known why she failed to arrive in New York. In the same way, had the equipment been developed a little sooner, then the fate of the *Mary Celeste* would *not* be a mystery. Ironically, the inventor himself had been offered free passage aboard the *Titanic*'s maiden voyage but had instead chosen to travel three days earlier on another ship. Back in the Marconi Station, an employee called David Sarnoff was coordinating the rescue efforts and listing the names of the known survivors. Apparently he alone manned the station for seventy-two hours without a break, or so he claimed, but this was not how Sarnoff would secure his place in wireless radio history. Sarnoff has an even better story than that.

For it was David Sarnoff, an ambitious Marconi employee, who realized there was a much greater potential for the use of wireless radio waves than simple point-to-point communication. The telephone had already been providing that service since 1892, albeit with the use of wires that limited its reach. Sarnoff, on the other hand, recognized that the same message could be picked up by multiple receivers, if they were all using the same radio-wave frequency. If he could have one listener, he reasoned, then why not one hundred, or one million, or even ten million, for exactly the same cost to the broadcasting company? But he had to be cautious as in 1913 an inventor called Lee de Forest (1873–1961), who worked at the Federal Telegraph Company, was being sued by the United States Federal Attorney on behalf of shareholders who felt they had been defrauded by his own plans to develop wireless radio. The Prosecuting Attorney is recorded claiming that, 'Lee de Forest has said in many newspapers and over his signature that it would be possible to transmit the human voice across the Atlantic before many years. Based on these absurd and deliberately misleading statements, the misguided public has been persuaded to purchase stock in his company.'

De Forest was later acquitted but was nearly bankrupted in the process. Sarnoff learned the lessons and, instead of making public announcements, he quietly experimented until he hit upon the idea of broadcasting music from a gramophone player. It was the first time that radio-wave technology had been thought of as a medium for entertainment, rather than for transmitting information. His colleagues were less than impressed and one famously commented, 'The wireless music box has no imaginable commercial value. Who would pay for a message sent to nobody in particular?' Undeterred, in 1916 Sarnoff outlined his ideas in a memo to Edward J. Nally,

Vice President and General Manager at Marconi, who, whilst recognizing the potential, deferred the idea as the company was already stretching its resources thanks to the ongoing First World War.

In 1919 the General Electric Company of America bought Marconi, and Sarnoff again submitted his memo, this time to Owen D. Young, the new Chief Executive who had formed the Radio Corporation of America (RCA), which had dealt primarily with military communications, during the same year. Again Sarnoff was ignored, but with the increase of amateur radio enthusiasts, using self-built receivers all across America, Sarnoff finally demonstrated the potential of his idea by arranging commentary of a heavyweight boxing match between the legendary Jack Dempsey and the French war hero Georges Carpentier on 2 July 1921. It was billed as the fight of the century and the first with million-dollar ticket sales, as nearly 100,000 people turned up to watch. Meanwhile, a staggering 300,000 people listened to Sarnoff's radio commentary on crackling, home-made receivers all across the country. By the end of that year the demand for home radio equipment had become so large that transmitting stations were popping up in every state. The radio industry had been born, despite the predictions of esteemed American inventor Thomas Edison who claimed, in 1922, that 'the radio craze will soon die out in time'. Sour grapes for Mr Edison? In modern times nearly 85 per cent of Americans still listen to the radio at some point each day, as do more than 90 per cent of all Europeans.

So, whatever happened to the Italian politician Pietro Lacava who had suggested Marconi was a lunatic as a twenty-year-old? Well, he went on to enjoy spells as the Minister for Trade and Industry and Minister for Finance in successive Italian governments. No wonder the Italians never achieved anything meaningful after the Renaissance. I thought it was because they were all too busy having sex and watching other people play football. Instead it seems to be because they had men like Lacava in charge. He died peacefully on Boxing Day in 1912, three years after the lunatic Marconi had been awarded a Nobel Prize for his work.

How Wrong Can You Be?

Oprah Winfrey has become one of the most successful, and powerful, women in television, across the world. But it was no easy ride for the celebrated talk show host. Her path to fame and fortune has meant overcoming a rough and sometimes abusive childhood and enduring many career setbacks, including once being released from her job as a television reporter because she was considered 'unfit for TV'.

THE TELESCOPE – AND WHY THEY LAUGHED AT GALILEO

As far as public records go, it was a Dutch-German spectacle maker called Hans Lippershey (1570–1619) who accidentally invented the telescope. He certainly filed the first patent for a contraption in 1608 after, apparently, noticing two children playing with lenses in his workshop and remarking that they could make a distant weathervane appear to be closer by looking at it through two lenses, of differing strength, held at a small distance apart. Other suggestions include the idea that he simply stole the design from a rival spectacle maker. Either way, his was the first patent for the device, which was filed on 2 October 1608 with the States General of the Netherlands. Later that month a small mention was made of Lippershey's patent at the end of a diplomatic report issued by the Ambassador to the Kingdom of Siam. As the report was distributed across Europe, some of the leading scientists and mathematicians of the age began to carry out their own experiments. These included Englishman Thomas Harriot (1560–1621; see 'The Potato'), a Venetian called Brother Paolo Sarpi (1552–1623) and a relatively unknown geometry teacher from the University of Padua called Galileo Galilei (1564–1642), who happened to be visiting Venice when the report arrived.

Galileo had first come to the attention of the scientific world in 1586 when he published a book on his design for a hydrostatic balance (weighing machine) and had then created the world's first accurate thermoscope (thermometer). In 1609 the man from the famous town of Pisa was the first to recognize the full potential of the telescope but also realized that spectacle lenses were simply not powerful enough if he was to achieve any real success with the new invention, which he considered to be of military value. Galileo then set about teaching himself the precise craft of lens making and had soon managed to increase the power of the instrument, now called a telescope (from the ancient Greek word *teleskopos*, meaning 'far seeing'), by up to ten times that of the naked eye. In August of 1609 he returned from his home in Padua to Venice where he invited dignitaries from the Senate to climb to the top of the bell tower at St Mark's Basilica. Here, he demonstrated how his new invention

7

was able to see ships out to sea a full two hours before they could be spotted with the naked eye.

The Doge of Venice (the Duke) Leonardo Donato (1536–1612) immediately realized the value of a device that could warn of a hostile and advancing navy several hours earlier than had been previously possible and commissioned the telescope for his own navy. He then awarded Galileo a job for life as a lecturer and doubled his salary. It is easy to imagine how that would have been enough of an achievement for a forty-five-year-old provincial lecturer but Galileo and his telescope were only just beginning a journey that would change civilization forever, creating unity and division, and ultimately destroying the man himself. On 7 January 1610, Galileo turned his telescope from the horizons towards the sky, and what he could see would change the world for all time.

Previously man's understanding of the universe had been from what could be seen with the naked eye, which was limited to the moon and the stars. The brightest of these stars appeared to move in different directions to the fixed orbits of the constellations and nobody had been able to explain why. The perceived belief of the day was that the earth sat at the centre of the universe and the sun, moon and stars all revolved around it (mainly because the Bible said so). It was also believed that all heavenly bodies were completely flawless as God had intended them to be. But when Galileo studied the moon through his telescope he could see craters, mountain ranges and valleys. This revealed the moon not to be perfect and that must have meant that Planet Earth was not unique, as had always been insisted by generations of the men of cloth.

He then turned his attention to one of the wandering bright stars that was known to the Romans as Jupiter. To the naked eye Jupiter looks like all the other stars but Galileo immediately concluded that it must be another planet similar to the one he was standing upon. It was another world. He also noted the four smaller stars changing their positions every night around Jupiter, which he realized must be moons in their own orbit around that planet. This obviously meant they did not rotate directly around Planet Earth, as our own moon does, and Galileo believed he had something explosive on his hands. Was there another world out there? The book he wrote about his discovery, which was rushed to print in only six weeks, called *The Starry Messenger*, turned Galileo into an overnight sensation. However, many scientists laughed at what they dismissed as a basic mis-understanding. Some, on the other hand, saw this as confirmation of Copernicus's theory, presented a century earlier, that the sun was at the centre of the universe and everything revolved around it. Many of them kept themselves very quiet, as they knew this theory would not be welcomed by the most powerful and dangerous group of people on their own planet, the Roman Catholic Church.

However, Galileo still wasn't finished and he next turned his attention to another of the wandering stars, known by the Romans as Venus. Through his telescope he was able to record the changing shape and size of the planet over a period of months. Week after week he watched as Venus transformed from large crescent shape into a small flat disc. Then, as the shadows crept back across the surface of the planet, it returned to a large crescent shape again. Immediately, Galileo recognized the importance of this; it could only mean that Venus was orbiting the sun and not the earth. This meant to Galileo that the earth was definitely not at the centre of the universe after all; the sun was. And this revelation would have far-reaching and serious consequences as it brought the astronomer into direct conflict with one of the central beliefs, and teachings, of the Roman Catholic Church. For centuries they had been preaching that God had placed mankind at the centre of the universe and Galileo's telescope presented the first serious challenge to this claim. It was to be the beginning of a conflict between science and religion that continues to exist in modern times, because the Bible clearly states that 'the sun rises and sets and returns to its place' and that the 'Lord set the earth on its foundations; it can never be moved'. Put in simple terms, it meant that if Galileo's telescopic observations were correct then the entire principles of Christianity could be undermined. And they were.

Space travel is utter bilge.
Dr Richard van der Riet Woolley,
UK space advisor to the government, in 1956
(*Sputnik 1* was launched the following year)

Imagine then how awkward this was for Galileo who, as a good Catholic, did not want to contradict the established beliefs of the Church. He also wasn't keen on being broken upon the wheel, boiled in oil or burned at the stake as other heretics had been for questioning the claims of the Vatican. Instead he tried to find some middle ground and, whilst defending the idea that the sun was at the centre of the universe, suggested that every passage of the Bible should not be taken literally.

Perhaps, he proposed, they were written about a 'different kind of movement of the earth and not its rotations'. Well aware of the treatment heretics had been receiving at the hands of the men in cloaks at the time, he travelled to Rome in an attempt to persuade Vatican officials not to ban his ideas and, instead, to embrace them. Galileo's life, by this time, was in real danger and in Rome a decree was issued ordering him to abandon his theories, which he promptly promised to do. For the next decade Galileo avoided the controversy but was encouraged to revisit his telescope-based revelations by the election of Cardinal Barberini as Pope Urban VIII in 1623. Barberini had previously been a friend and supporter of Galileo.

However, his 1632 book, *Dialogue Concerning the Two Chief Systems*, which was intended to be a balanced account of the diametrically opposed views, made up of healthy debate, was met by a summons to trial by the Inquisition in Rome where he was immediately charged. By this time Galileo must have been wishing he had never made his telescope as, despite his repeated denials that he was rejecting the scriptures, he was found guilty of heresy: 'namely of having held the opinions that the Sun lies motionless at the centre of the universe, that the Earth is not at its centre and moves, and that one may hold and defend an opinion as probable after it has been declared contrary to Holy Scripture'. Under threat of even worse punishment, he was instead sentenced to a life of house imprisonment and his works were banned from circulation. He was also ordered to read the seven penitential psalms once a week for the following three years.

Galileo died in January 1642 – one of the few who questioned the Roman Catholic Church's Holy Scriptures to do so peacefully. His telescope, which was invented as an effective military tool, remains one of the few inventions that has literally changed the course of the history of mankind and stands as the provider of one of the first, and key, pieces of evidence that the Catholic Church were not to be trusted.

AIR CONDITIONING: THE KING OF COOL

Nobody is sure of the exact date of the beginning of mankind. Many historians estimate that the first evidence of controlled fire can be dated to one million years ago, whilst others point to the evidence of cooked food as far back as 1.9 million years ago. One thing we can all assume is that humans could not have evolved without fire and so the two are irrevocably connected.

And it is fire that has been used to enable humans to live in cold climates from the dawn of the species until now, although, admittedly, it is not the only option these days. Two thousand years ago the Romans, who had already mastered the art of underfloor heating, made attempts at cooling a room by circulating cold water from aqueducts around their buildings. A few centuries later the Chinese invented the rotary fan, which was to remain the most effective way of keeping people cool for the next 1,700 years.

Efforts were undertaken to make the long summers living in the new big American cities more comfortable, and in 1758 Benjamin Franklin and John Hadley, a Cambridge University professor, experimented with the liquid evaporation of alcohol and other volatile liquids that were found to cool an object down far enough to be able to freeze water. But, up until the late nineteenth century, the idea of having complete control of the

12

environment inside a building was as ridiculous as the idea of halting the rain or preventing the sun from shining.

However, in 1901 Willis Carrier, a young electrical engineering student, graduated from Cornell University and was offered a position at the Buffalo Forge Company in Buffalo, New York, which specialized in making electric fans. His first assignment was with the Sackett & Wilhelms Lithography Company of Brooklyn, which was responsible for producing *Judge*, America's most popular full-colour magazine at that time.

To place a man in a multi-stage rocket and project him into the controlling gravitational field of the moon, where the passengers can make scientific observations, perhaps land alive, and then return to earth? All of that constitutes a wild dream worthy of Jules Verne. I am bold enough to say that such a man-made voyage will never occur regardless of all future advances.

Sir Harold Spencer Jones, Astronomer Royal of
the United Kingdom, in 1957

But in the heat of July 1902, Sackett & Wilhelms had a big problem. Because of the humidity levels in their buildings, the new colour ink they were using for their front covers would not stick to the page and was simply sliding off. Carrier was tasked with finding a way of reducing the temperature in the printing room to precisely 53°F, and the humidity as quickly as possible or the printer faced the reality of failing to deliver the monthly issue to their millions of subscribers.

Carrier worked night and day on the problem and early one morning, after a few weeks had passed, he was standing on the platform of a foggy railway station when he made his life-changing breakthrough. The engineer knew that fog was simply air that was saturated with water and realized that if he could create 100 per cent humidity in a room he would then be able to introduce enough dry air to reduce humidity by precise levels. He would be able to reproduce whatever level of humidity he needed.

That day was 17 July. Carrier faced a race against time if the deadline for the August edition was to be successfully met so he immediately began to work on his theory. He already understood how to heat objects with steam by passing air through hot coils and his simple plan was to reverse the process by pumping air through coils that were cooled down by water. With the use of his fans, he found that he was able to control the temperature, humidity, air circulation and ventilation. The low heat and humidity he immediately achieved in the print room helped the paper to remain consistent and enabled perfect ink alignment; the American public was not to be deprived of its favourite magazine the following month.

By 1907 Carrier had improved his design and introduced what would later be called by air-conditioning engineers the Law of Constant Dew-Point Depression; he applied for a patent on 17 May, which would eventually be issued on 3 February 1914. However, war in Europe led to manufacturing companies concentrating their efforts in other areas so Carrier decided to leave the Buffalo Forge Company and, with six other young engineers, formed the Carrier Engineering Company of New York. Demand for their air-conditioning units remained steady, although it wasn't until 1925, when they designed an air-conditioning system for the Rivoli Theatre on Broadway, that the general public would be introduced to controlled indoor environments for the first time. Such was its success that the Rivoli found itself packed to the rafters at all times of day or night as New Yorkers sought relief from the steaming city summer, whatever movie it was they were showing.

Sadly, the stock market crash of 1929 and the Great Depression that followed slowed the company's progress, although by 1937 it had still become the largest employer in central New York. But it was during the post-war economic boom of the late 1940s and 1950s that Carrier's invention began to revolutionize America. Cinemas, restaurants, factories, schools, hospitals, public buildings, shopping malls and any developing city in the Midwest, West Coast or the Deep South were installing Carrier's air-conditioning units as fast as they could be produced.

Millions of East-Coasters were now prepared to move to the previously inhospitable west and south of the United States thanks to the new air-conditioned environments, and both economic and political power shifted with them. Where the East Coast communities once dominated the American way of life, now cities such as Dallas, Phoenix, Atlanta, Miami and Los Angeles grew into the country's most powerful urban centres. Only New York, Chicago and Philadelphia from the old East Coast guard remain among America's largest ten cities. All thanks to the air-conditioning unit.

Throughout his life Carrier received many awards and accolades for his invention – an invention that changed the way mankind exists. Can you imagine, today, surviving in some of the climates of South East Asia or Africa if a twenty-five-year-old engineer had not been tasked with reducing the humidity in a printing room and not had that moment of inspiration on a foggy New York platform that proved something generally considered to be impossible was, in the end, easily achievable? Willis Haviland Carrier died a wealthy man in 1950 and the company bearing his name lives on with annual sales in excess of $15 billion and a staff of over 45,000 people in useful, and comfortable, employment.

> **How Wrong Can You Be?**
> As a boy Charles Darwin abandoned all ideas of having a medical career and was often criticized by his father for being a 'lazy dreamer'. Darwin later admitted, 'I was considered by all of my masters, and my father, to be a very ordinary boy and rather below the common standard of intellect.' Darwin later wrote, in the foreword to his ground-breaking book *The Origin of Species* (1869), 'I see no good reason why the views given in this volume should shock the religious sensibilities of anyone.'

THE ROBOTIC ARM

American inventor George Charles Devol, Jr (1912–2011) was eleven years old when Karel Čapek's influential play *R.U.R* (Rossum's Universal Robots) first appeared at New York's Garrick Theatre in October 1922 and captured the imagination of the American public. The word 'robot' was suggested by Čapek's brother as it translates as 'serf labour' or 'hard work' in the modern Czech language, and robots were soon appearing in film, books, comics, radio plays and in just about every other form of entertainment across America. The young George would surely have noticed as he showed an early interest in all things mechanical and electrical. In 1932 Devol decided against further education and instead formed his own company, United Cinephone, and, rather than competing with the electrical giants of the day, chose to start inventing his own products.

Among his early achievements was the automatic electric door (now a feature of just about every public building in the world) and a system for sorting packages that would lead to the modern barcode. It is difficult to imagine that nobody before Devol had ever thought of opening a door without using a handle and either pushing or pulling. United Cinephone would also file patents for lighting, packing, printing and automatic laundry-press machines. During the Second World War Devol sold his company and began working on radio, radar and microwave technologies, and was involved with the project that supplied counter-radar systems to all Allied aircraft on D-Day and beyond. This, of course, was the forerunner to the modern-day, invisible Stealth Bomber. But it was after the war that Devol was to make his greatest contribution to modern life.

George Devol's favourite place was his garage in Connecticut where he would spend most of his time thinking, puzzling and looking for new and inventive ideas. One day, whilst reading a technical magazine, he saw a photograph of an assembly line and began to wonder why human beings should have to do such repetitive and boring work that required nothing more than recurring arm movements. He realized that by doing such mindless tasks human beings naturally lose concentration, which leads to injury. He also knew that if he could come up with a machine to do all the repetitive work then conditions in the workplace could be improved. He quickly understood that if he could invent a tool that moved in a similar way to the human arm, and could hold onto things and place them with precision, then its potential uses would be endless.

> Professor Goddard with his 'chair' in Clarke College and the countenancing of the Smithsonian Institute does not know the relation of action to reaction and of the need to have something better than a vacuum against which to react. To say that would be absurd. Of course, he only seems to lack the knowledge ladled out daily in our high schools.
>
> *New York Times* editorial about Robert Goddard's rocket technology in 1921, retracted by the newspaper on 17 July 1969

George immediately sketched a device with a wrist that could move independently to the arm, with two opposing fingers that would form a strong grip. The arm's movement would be controlled by a computer that could be programmed to do thousands of different tasks. Although he obtained a patent for his invention Devol was unable to find anybody in American industry to listen to his proposals and was repeatedly told his product was a bad idea that nobody would ever invest in. Doors closed all along the path ahead. Several years later, George was at a friend's party when he was introduced to a fellow guest called Joe Engelberger, who was chief of engineering at an aviation company. The two men sat down with a drink and before long Devol was explaining and sketching out his idea for the robotic arm and, for the first time, noticed he was speaking to somebody who fully understood his invention. Engelberger, for his part, sat in amazement. He immediately knew that what was being explained to him was something he had only read about in science-fiction magazines and that his new friend understood how to actually make it work.

Unlike Devol, Engelberger was an entrepreneurial businessman who recognized the number of potential uses the robotic arm would have, and it was the start of a world-changing relationship. As Devol set about building a working prototype Engelberger toured the factories of his industry colleagues looking for ways the machine could make the biggest difference as quickly as possible. In other words, he was searching for ways to make an instant impact by doing the important jobs that humans did not want to do. Or the dangerous ones.

When the first prototype, the Unimate, was ready, George and Joe tested it on as many mundane tasks as they could think of and found that it could be reliably programmed to perform just about any assignment they gave it. They were certain of great success but quickly realized that the image robots had in America was anything but positive. At that time robots were always the bad guys and in the movies they could usually be found roaming around causing havoc and killing people. They were the stuff of horror movies, and the lack of enthusiasm for their invention proved to be a big surprise. Engelberger suggested they expand their horizons and so, after mass rejection in America, he turned his attention to Japan, where the post-war economy was also booming.

Within months several Japanese car manufacturers were using the Unimate on their assembly lines and found they had a new worker that performed twenty-four hours a day, never took a break, never had a holiday, never picked up an injury and, most importantly of all, never complained. Productivity soared and Japanese car

19

manufacturers began to dominate the industry with their reliable and well-made products. The rest of the world now had some catching up to do as they still had a man with a spanner working on an eight-hour shift competing with Devol's robotic arm.

Within a few years industries throughout the world were computerizing their production lines and the robotic arm began to be employed in hundreds of thousands of new ways, saving thousands of lives in the process, including on the US space programme and throughout bomb-disposal teams of every nation's army. Devol and Engelberger had been vindicated, just as they knew they would be from the beginning. *Popular Mechanics* magazine even listed the Unimate as one of the top fifty inventions of the century.

THE X-RAY IS A HOAX

Like many of the great medical innovations and discoveries, the X-ray was happened upon by chance. Wilhelm Röntgen (1845–1923) was a German physicist who studied mechanical engineering at the University of Zurich and was appointed Professor of Agriculture at the University of Wurzburg in 1875. On 8 November of that year Röntgen was carrying out a series of experiments relating to cathode rays, which were first identified in 1869 by another German scientist, Johann Hittorf, but were still virtually unknown at that time.

Röntgen was studying the external effects of these rays when they were passed through a vacuum in a glass tube and noticed that any fluorescent surface nearby would become luminous, even when shielded from the direct light. He then noted that when a thick metal plate was placed between the tube and the fluorescent surface a dark shadow was cast, and when replaced by an object of less density, such as his jacket, a much weaker shadow was observed. He also recorded that the invisible rays caused cardboard and other dense material to appear fluorescent.

Röntgen was bemused and, late in the afternoon, decided to build a black cardboard box to completely cover the tube. After turning off all the lights in his laboratory, he began a series of experiments. Each time, he noticed a faint shimmering of light appear about one metre away. He struck a match to investigate and discovered the light had been emitted from an unrelated screen filled with barium platinocyanide that he had been intending to use for something else entirely. Once again Röntgen began placing objects of various densities between the light and his cathode rays and noticed nothing unusual, except that the image of his hand appeared on the screen without any flesh, the bones clearly identifiable.

Still unsure of exactly what type of ray was causing this new phenomenon, the scientist named his new discovery by using the mathematical term commonly applied at the time for anything unknown: 'X'. Within two weeks Röntgen had taken the first ever X-strahl (ray) of his wife's hand who, upon seeing her own skeleton, became faint and declared, 'I have just seen my own death.' Very soon the scientific community were calling this new discovery 'Röntgen-Rays', as they are still known in many countries, although the man himself disapproved and insisted his important discovery be known only as X-rays.

However, it was a long time before some members of the scientific community were convinced, and twenty-four years later, in 1899, the eminent Scottish scientist William

Thomson, Lord Kelvin, declared that 'X-rays will prove to be a hoax'. They never were. Mind you, this is the same William Thomson who strongly disputed Charles Darwin's Theory of Evolution, publicly declared that 'radio had no future' and announced that 'heavier than air flying machines are impossible'.

> I am tired of all this sort of thing called science here. We have spent millions on it in the last few years and it is time it should be stopped.
>
> Simon Cameron, US Senator,
> on the Smithsonian Institute in 1901

THE TELEPHONE IS A MEANINGLESS TOY

During the early 1800s several attempts were made to connect the East Coast of America and the towns and cities that were establishing themselves at the western frontier as European immigrants spread out. From 1828 the Great Western railroads had been delivering cargo, supplies and mail, but prior to then the stagecoaches were the only means of communication, and that could take months. Things changed in 1838 when Samuel Morse invented the first reliable electronic telegraph that was able to transmit messages, using a code he also invented bearing his name, over long distances that could be received almost instantly. The telegraph was an immediate success and over the following two decades electronic wires were strung up all over the country, sometimes from wooden poles and sometimes from trees. As the trees swayed on the breeze the wires would stretch and curl, leaving them with the appearance of a grape-vine, which is how the telegraph earned its affectionate

22

nickname and became an established part of the English language.

However, by the 1860s a new generation of electrical engineers was experimenting on ways to transmit a voice along the wires, and in 1876 two of them, American Elisha Gray (1835–1901) and British engineer Alexander Graham Bell (1847–1922), filed patents at the US Patent Office in New York on the very same day, 14 February. Bell won the subsequent dispute because his patent had been submitted by his lawyer two hours before Gray's had and so earned his place in history. And what happened to Gray, on the other hand? Well, imagine if he later discovered that his lawyer had stopped off for a spot of lunch on his way to the Patent Office. On such a fine thread hangs the balance between immortality and anonymity, fame and obscurity.

The dispute became public and was met with general indifference within the communications and telegraph industry. An internal memo at the Western Union Telegraph Company of New York, during 1876, was published that revealed, 'This "telephone" has too many shortcomings to be seriously considered as a means of communication. The device is inherently of no value to us.' The President of Western Union, William Orton, was convinced that the telegraph had already become the 'central nervous system of commerce', and could not be replaced. The British were even less enthusiastic, which was revealed when Sir William Preece, the Chief Engineer of the Post Office, loftily announced, 'The Americans may have need of the telephone, but we do not. We have plenty of messenger boys.' (Preece learned his lesson about embracing new technology – see 'The Radio'.)

23

In 1868 the *New York Times* ran a news item announcing that 'a man has been arrested in New York for attempting to extort funds from ignorant and superstitious people by exhibiting a device which he claims can convey a human voice over any distance through metallic wires so that it can be heard by the listener at the other end. He calls this instrument a telephone. Well-informed people know that it is impossible to transmit the human voice over wires.'

The *Boston Globe* also ran an article that included the statement, 'Well-informed people know that it's impossible to transmit the human voice over wires as may be done with the dots and dashes of Morse code, and that, were it possible to do so, the thing would be of no practical value.'

Despite this negative reaction to his invention, Bell and his team of engineers continued to work on the idea and in August 1876, for the first time in history, a voice could be heard along the wire from a distance of six miles. But the US President Rutherford B. Hayes, after being given a demonstration, noted, 'It's a great invention, but who would want to use it?' Bell and his financial backers, Gardiner Greene Hubbard (who became Bell's father-in-law the following year) and Thomas Sanders, then offered the patent to Western Union for $100,000, and again William Orton responded that the telephone was 'nothing more than a meaningless toy'.

Only two years went by before Orton is known to have claimed to colleagues that if he could 'now buy the patent for $25 million I would consider it a bargain'. But his chance had gone and the Bell Company no longer wanted to sell. By 1886 over 150,000 Americans had telephones, and Bell, Hubbard and Sanders were very wealthy men. Against the considered advice of the known experts of their day, the Bell Company created an industry that is today estimated to be worth around $5 trillion per year.

How Wrong Can You Be?
Albert Einstein was unable to speak until he was four years old and failed to read a word until he was nearly eight. His parents and his teachers began to believe he was either mentally handicapped or simply anti-social. Finally, he was expelled from school and denied a place at the Zurich Polytechnic. However, it is generally agreed that he caught up in the end when he won the Nobel Prize in Physics and completely altered many commonly held scientific beliefs during his lifetime.

COMPUTERS – WHO NEEDS THEM?

In 1924 George W. Fairchild, Chairman of the Computing Tabulating Recording Company (CTR), died and was succeeded by the head of the sales team and company President, Thomas J. Watson. CTR's product range at the time included cash registers, weighing machines, meat slicers, punch-card equipment, adding machines and time-keeping systems. As soon as Watson was in sole control he changed the name of the company to International Business Machines (IBM), which was ambitious as the firm did not even have a national reach at the time. However, within four years Watson had doubled the company's revenue to $9 million.

Business grew rapidly and by the 1930s IBM's German subsidiary was the most profitable in the company, largely thanks to the punch-card machines provided to the Nazi Party and their tabulation of census data that recorded the location of citizens by race, gender and religion (i.e. where the Jewish people were living). For this Watson was awarded the Order of the German Eagle in 1937, although when he returned it in 1940 Adolf Hitler was said to be furious and declared that Watson would never set foot upon German soil again.

During the war IBM continued to trade profitably, although when considering new emerging technologies and products Watson famously announced in 1943 that 'I think there is a world market for maybe five computers, maximum.' Watson continually resisted any involvement with electronic computing until his retirement in 1949, regarding it as expensive and unreliable, even after his dismissive prediction had already proved to be inaccurate. His son and successor, Tom Jr, had other ideas and immediately began hiring electrical engineers to design and build mainframe computers; in 1950 the custom-designed US Air Force SAGE tracking system was responsible for over half of IBM's computer sales that year.

Despite many recognizing the potential of the new technology, the fact remained that there were still only twelve mainframe computers in the world. Watson Jr employed an expert to determine if there was a market for computers and Cuthbert Hurd, of the Atomic Energy Commission's Oak Hill National Laboratory, predicted that 'he could find customers for about thirty machines'. Watson fired him on the spot. Well, if he didn't then he should have done. IBM's computer sales failed to make any profit in 1950, although Watson Jr later said, 'but it enabled us to build highly automated factories ahead of anybody else and to train thousands of new workers in electronics'. His cool reserve paid off handsomely as by 1956 Watson Jr had tripled IBM's income from $200 million to $743 million and continued that rate of growth throughout his twenty years as Chairman, as IBM dominated the world's growing computer industry.

Watson Sr's lack of vision did not end with personal computers. Between 1939 and 1944 IBM was one of twenty companies who would turn down Chester Carlson's design for an electric copying machine. At the time the only way to make copies of any document was to use a sheet, or multiple sheets, of carbon placed between paper pages and the technical innovators of the day did not see any reason to try to improve upon that system. In 1949 IBM even released a lofty statement that concluded that, 'The world potential market for copying machines is five thousand at the most. It has no market large enough to justify production.' As we know, they had been wrong before.

> Animals, which move, have limbs and muscles. The earth does not have limbs and muscles and therefore it does not move.
>
> > Scipio Chiaramonti (1565–1652), Professor of Philosophy and Mathematics, University of Pisa, dismissing Galileo's theory in 1633

As it turns out, 1949 was the very year Carlson, a clerk in the New York Patent Office, refined his machine and renamed the process 'xerography', which he adapted from the Greek word for 'dry writing'. He formed the

Xerox Corporation and made himself a $150 million fortune, most of which he gave away to children's charities after telling his wife that his last ambition was to 'die a poor man'. Carlson's fortune, however, would not be the sum of IBM's loss in rejecting his machine. In 1955 he sold the rights to the Xerox machine for what would eventually be calculated as being worth one-sixteenth of a cent for every Xerox copy made anywhere in the world. You do the math.

Between the 1960s and the 1990s, Digital Equipment Corporation (DEC) provided a credible challenge to IBM's dominance of the computer industry and millions of sales of their PDP range led them to become the second largest company of its kind by the mid 1980s. However, DEC failed to recognize the rapid rise of the microcomputer during the 1980s and the parallel demand for the home computer. In 1977 company founder Ken Olsen revealed he was looking the wrong way when he announced, 'There is no reason why anyone would want a computer in their home.' This is despite having one in his own home.

The rise of the home computer and the use of word processors in the workplace have led to sales of billions of units around the world over the last twenty years but during that time DEC's decline was as spectacular as its emergence three decades earlier. In 1992 Ken Olsen was replaced as the company's president by Robert Palmer who began a programme of downsizing as he struggled to keep the firm alive. Eventually, in 1998, DEC was sold to Compaq and its PC manufacturing division was quietly closed down. IBM, on the other hand, which led the way into people's homes and offices with the IBM Personal Computer (IBM PC) in August 1981, remains one of the largest technology companies in the world with an estimated value of $214 billion.

THE JET ENGINE
In 1916, as the First World War was well underway and short steps were being taken in the development of aircraft, a monumental event took place that would, eventually, change the course of aviation forever. And that was the emergency landing of a single-seater aircraft with engine trouble near the town of Royal Leamington Spa in Warwickshire, England. Now, ordinarily there would be nothing significant or even unusual about this, but the minor accident was witnessed by a nine-year-old schoolboy called Frank Whittle (1907–95), the son of a local engineer who owned the Leamington Valve and Piston Ring Company. By then the young Whittle was already showing a keen interest in engineering and would soon become an expert on his father's single-cylinder gas engine. And he was immediately fascinated by what he saw that afternoon. In fact, he became so absorbed by the flying machine that he failed to notice when the pilot was preparing to take off again, and he was almost run down.

Frank Whittle then spent the next five years developing his mechanical awareness in the workshop and studying the theory of flight, astronomy, engineering and, crucially, turbines at the local Leamington library. By the time he was fifteen years old Frank was determined to become a pilot himself and in 1922 he applied to join the Royal Air Force. In January 1923 Whittle breezed through the entrance examination but was halted at the second hurdle when he failed his medical on account of his height; he was only five feet tall. His chest was also determined to be 'too small'. In an early display of dogged tenacity Whittle then spent the next six months with a physical training instructor, increasing his chest size, and, by the time he reapplied, he was both three inches taller and wider. But he was told that candidates were not allowed to submit a second application and was turned away.

Even then Whittle refused to be defeated and applied again, this time under an assumed name, and was finally accepted on a three-year training course as an apprentice mechanic at RAF Cranwell in Lincolnshire. At the third time of asking, Frank Whittle was finally enrolled at the RAF's Technical School of Training and this determination would become a feature of his working life.

Frank soon regretted his decision as his rebellious nature was frequently at odds with the strict regime of discipline within the armed forces. He also managed to convince himself that he had no chance of ever becoming a pilot. Whittle considered deserting but his involvement in the Model Aircraft Society, and the quality of his working replicas, attracted the attention of the commanding officer, who recommended the young engineer for officer training. Whittle immediately recognized this as his big opportunity; the course included flying lessons and by 1927, after only fourteen hours in the aircraft, he was cleared to fly solo. This was when Whittle decided to make a career of the RAF and also, for his graduation thesis, chose to write a paper on potential aircraft-design developments that included flying at high altitude and at speeds of over 500 mph. Both objectives had been considered before and dismissed as 'unachievable' at the time. The twenty-one-year-old Whittle, however, had other ideas. His paper entitled 'Future

Developments in Aircraft Design' argued how unlikely it was that those speeds could be reached by using conventional piston and propeller engines: instead he suggested an alternative – turbine engines – and his final paper won him the Abdy Gerrard Fellowes Memorial Prize for Aeronautical Science. Whittle was also described by examiners as 'an above average to exceptional pilot'.

In 1929 Pilot Officer Whittle was posted to the Central Flying School on the south coast of England where he became an instructor. Whilst there, the young engineer showed his designs and proposals for a 'turbine' or 'turbo-jet' engine to Flying Officer Pat Johnson who had previously been a patent examiner. Johnson then briefed their commanding officer and Whittle was encouraged to submit his design to the Air Ministry, who failed to grasp the concept of the jet-propelled engine themselves and passed the papers to A. A. Griffith, a senior engineer whom they knew to be working on a similar idea. Whittle was invited to a meeting by the curious Griffith where he explained his design and expressed how he 'totally believed in it'. However, Griffith remained unimpressed and told the would-be inventor, 'I am sure you do, but it is totally impractical.' He went on to outline what he described as 'fundamental flaws' in the calculations and Whittle was then sent packing. Griffith's subsequent report to the top brass led them to completely reject Whittle's proposal.

Needless to say, Whittle and Johnson remained optimistic and went ahead to file a patent for the design in January 1930. Since the RAF had dismissed the concept there was no need to declare it a secret and Whittle retained the commercial rights. When his patent was published the German Trade Commission in London obtained a copy and immediately sent it to the German Air Ministry and the German aero-engine manufacturers, who pored over the details. Then in 1935 Whittle received a letter from a fellow officer cadet, Rolf Dudley Williams, who proposed raising funds to actually develop a jet engine themselves. Whittle agreed and Williams then introduced two other friends, J. C. B. Tinling and Lancelot Law Whyte of the investment bankers O. T. Falk & Partners. The four men then agreed to form Power Jets Ltd in November 1935. Meanwhile, the Germans and their own engineer, Hans von Ohain, began working on a similar idea but it was Whittle who fired up the first working prototype in 1937. It was not without its teething problems and the team nearly panicked when they found the turbine kept accelerating even after they had turned it off. They later discovered that fuel had leaked and gathered around the intake. The engine finally began to slow once that had burned away.

> There is a young madman proposing to light the streets of London. With what do you suppose, smoke?
> Sir Walter Scott (1771–1832) on hearing proposals
> to light the streets with gas lamps

Once again Frank Whittle was unable to attract the attention of the Air Ministry and his patent lapsed after two years because he could not afford the £5 renewal fee. The RAF had refused to reimburse him. He was then sent to Cambridge University to study for a Mechanical Science degree, where an aeronautics professor took one look at the design before remarking, 'Yes, very interesting Whittle dear boy, but it will never work.' The Germans, on the other hand, were far more convinced and they too had their own working prototype fired up by the end of 1937. With the benefit of hindsight it is easy to see why the German Air Ministry were more serious about developing the jet engine, as they were gearing up for war in Europe. The British, however, were still refusing to recognize such a prospect and the investment partners of Power Jets Ltd were left to fund the development to the tune of only £2,000 with a promise of a further £18,000 over a period of eighteen months. Whittle, a serving RAF officer, was given special permission to work on the project on the condition that it was for no longer than six hours a week – a clear indication of how little value the British government and Air Ministry placed upon his jet-engine project.

Meanwhile, over in Nazi Germany, Hans von Ohain was ready to demonstrate his actual flying prototype in 1939. As war broke out in September of that year, Power Jets Ltd still had a staff of only ten. The prospect of a total European war ahead, coupled with Whittle's frustration at the Air Ministry, began to take its toll on the engineer. Whittle wrote: 'The responsibility that rests on my shoulders is very heavy indeed. Either we place a powerful new weapon in the hands of the Royal Air Force or, if we fail to get our results in time, we may have falsely raised hopes and caused action to be taken which may deprive the Royal Air Force of hundreds of conventional aircraft that it badly needs. I have a good crowd round me. They are all working like slaves, so much so that there is a risk of mistakes through physical and mental fatigue.'

For his own part Whittle began to suffer from stress, which led to heart palpitations and eczema. His weight dropped to only 57 kg and he began to rely on Benzedrine to help him through the sixteen-hour days and tranquilizers in order to sleep at night. By the time the Air Ministry paid yet another visit to Power Jets, who barely had enough money to keep the lights on at their workshop. Whittle and his team had managed to run the jet for twenty minutes without any problems, and a ministry official, the Director of Scientific Research, David Randall Pye, left the demonstration finally convinced of the engine's potential. In January 1940 the Air Ministry placed an order for a single test engine and within three months had issued a contract for Rover cars to establish a production line that would deliver up to 3,000 jet engines per month within two years. Frank Whittle was then promoted to the rank of Wing Commander.

On 15 May 1941 the first jet engine was demonstrated in test flight and was a great success, although neither the Germans nor the Allied Forces were able to mass produce fighter-jet aircraft in time for them to have any meaningful effect on the outcome of the war. However, it could be argued that if the British government had taken Frank Whittle and his jet engine design more seriously during the mid-1930s then the jet would have had a significant role to play in bringing the Second World War to a close much sooner than it did eventually end.

Finally, after nearly twenty years, Frank Whittle had been vindicated and recognized for his revolutionary and life-changing vision. Before long the Americans had latched onto Whittle's jet engine and they quickly understood how it could transform transatlantic travel during peacetime. The jet set, as they became known, were soon leaving London and, thanks to the five-hour time difference, were able to arrive in New York only a few hours later – a journey that previously had taken up to a week on the fastest ships. In July 1943 Frank Whittle was promoted to the rank of Group Captain, and the following January he was awarded a CBE in recognition of his work. In April 1944 the British government decided to nationalize his company and Whittle was paid only £10,000 for his shares. But, by then, he had already been admitted to hospital where he spent six months recovering from nervous exhaustion. After he had been discharged Whittle, a lifelong socialist up until that point, changed his political allegiances and supported the Conservative Party for the rest of his days.

In May 1948 Frank Whittle was finally awarded £100,000 from the Royal Commission in recognition of his work and two months later he was made a Knight Commander of the Order of the British Empire. Unfortunately, within a few months, whilst on a gruelling lecture tour of the United States, Whittle once again broke down, leading to his retirement from the Royal Air Force on medical grounds in August of that year.

After he recovered for a second time Frank spent the rest of his working life as a technical advisor for various international companies including BOAC and Shell Oil, breaking only to work on a biography called, appropriately, *Jet: The Story of a Pioneer*. He then emigrated to the United States after accepting the position of NAVAIR Research Professor at the US Naval Academy in Maryland. Frank Whittle, the man whose single-minded and determined vision gave us the modern jet passenger liner, died peacefully on 9 August 1996 and his ashes were returned to England, where they were placed in the RAF Chapel at Westminster Abbey, London.

How Wrong Can You Be?

In September 1971 the outspoken American biologist and author of *The Population Bomb* (1968) Dr Paul Ehrlich gave a speech to the British Institute for Biology, during which he claimed: 'By the year 2000 the United Kingdom will be simply a small group of impoverished islands inhabited by seventy million hungry people. If I were a gambler I would take even money that England will not exist in the year 2000 AD.' He also claimed that India would be 'unable to feed more than two-hundred million people by 1980'. The *New Scientist* magazine later endorsed the speech in an editorial entitled 'In Praise of Prophets'. Ehrlich also predicted that sixty-five million Americans would starve to death during the 1980s and that by the end of the millennium the population of the United States would have dropped to only 22.6 million.

SATELLITE COMMUNICATIONS

A satellite is defined as either a 'celestial body orbiting another, larger, body or a man-made object (vehicle) designed to orbit the earth, moon or other celestial body'. The idea for today's modern satellite communication network was first conceived during the aftermath of the Second World War as a means of tracking and monitoring radio signals being transmitted between the Soviet Union and Eastern Europe. Essentially a device for spying on the Russkis, the first communications satellite was, in fact, launched into orbit by the Soviets themselves on 4 October 1957, at the start of the Sputnik programme, which initiated the fabled Space Race of the 1960s with the United States of America. The Americans, who were taken by surprise, had announced their intention to launch communications satellites as early as 1955 and were stunned to find the Soviets had stolen a march on them. However, they responded on 31 January 1958 when their own artificial satellite, Explorer 1, was launched.

But the idea of satellite communication was not a new one in 1957 by a very long way. As early as 1728 the English physicist and mathematician Isaac Newton (1642–1727), who was considering the forces of gravity and planetary motion, had outlined his theories in a book, *A Treatise of the System of the World*, published the year after his death, which first considered the possibility of orbital satellites. Then, in 1879, the science-fiction author Jules Verne (1828–1905) described artificial satellites in his popular novel *The Begum's Millions*. In 1903 the Russian scientist Konstantin Tsiolkovsky (1857–1935) released the first academic study on the use of rockets to launch spacecraft in his book *Exploring Space Using Jet Propulsion Devices*, and twenty-five years later Herman Potoènik (1892–1929) was describing the use of orbital spacecraft for both peaceful and military observations. He also suggested the use of radio technology as a means of communicating between the craft and Planet Earth.

But the idea was addressed seriously for the first time by another science-fiction writer, Arthur C. Clarke (1917–2008), who wrote an article for *Wireless World* in October 1945 called 'Extra Terrestrial Relays – Can Rocket Stations Give World-Wide Radio Coverage?' This examined the possibility of a network of orbiting satellites enveloping the entire planet that would enable high-speed global communications. The US military then began to take a closer look at the available technology and the following year, in May 1946, Project Rand, which had been established with the objective of studying the long-term future of military weapons, released its paper entitled 'Preliminary Design of an Experimental World-Circling Spaceship', which it described as 'one of the most potent scientific tools of the Twentieth Century'. However, not everybody was quite as enthusiastic. The United States Air Force released a statement explaining that it did not believe that the satellite had any military potential and dismissed it as a 'tool for science, politics and propaganda'. And it had not been the only one to flatly dismiss the idea of communications satellites.

On 13 January 1920 the *New York Times* magazine published an article in which the editor predicted that 'a rocket will never be able to leave the earth's atmosphere' in response to discussions about Konstantin Tsiolkovsky's theory and in 1926 the American inventor Lee de Forest questioned the ability 'to place a man in a multi-stage rocket and project him into the controlling gravitational field of the moon where he can make scientific observations, perhaps land alive and then return to earth? All that constitutes a wild dream worthy of a Jules Verne novel. I am bold enough to say that such a man-made voyage will never occur regardless of any future advances.' He was not alone either, because even in 1961, three years after the first successful satellite launch, T. A. M. Craven, the former Federal Communications Commission (FCC) Commissioner, loftily announced that 'there is practically no chance communication space satellites will be used to provide better telegraph, telephone, television or radio services inside the United States'. Such was the general negativity about the whole idea that even after US Secretary of Defense James Forrestal announced on 29 December 1948 that his department was coordinating efforts that would lead to satellites being launched in the spring of 1958, his successor simply claimed, 'I know of no American satellite program.' There may, of course, have been an element of secrecy and subterfuge on the part of the American authorities during the period that became known as the Cold War.

Despite such denials, and the negative predictions of some of the finest minds of successive generations, the first satellite television broadcast was made in 1964 when coverage of the Summer Olympics, being hosted by Japan, was beamed into homes across America. This was a mere three years after the US's most senior government communications engineer had insisted that satellites could not improve global communications. As a result of Arthur C. Clarke's 1945 article in *Wireless World,* the science-fiction author is considered by many to be the inventor of the communications satellite. In reality it was the work of the US Navy and its Communication Moon Relay Project, which developed reliable technology that used the moon as a natural satellite after detecting radio waves that bounced off its surface. Its ambitious EME (Earth-Moon-Earth) Communications Project was clearly the forerunner of today's modern satellite communications network that was first imagined, in detail, by Arthur C. Clarke.

The machine gun is a much overrated weapon. Two per battalion is more than sufficient.

General Douglas Haig, 1915

In 1963 Clarke's vision won him the Franklyn Institute's Stuart Ballantine Medal and he twice served as the chairman of the British Interplanetary Society. During the 1960s Clarke became known as one of the world's leading science-fiction writers and produced numerous fiction and non-fiction books, including the seminal *2001: A Space Odyssey*, which would make him world famous. In an interview given shortly before his death in March 2008 Clarke was asked if he had realized satellite communications would become so important to the world. He cryptically replied, 'I am often asked why I didn't patent the idea of communications satellites. My answer is always "a patent is really a license to be sued".' Clarke was knighted in May 2000 for services to literature. The geostationary orbit for communications and weather satellites is informally known as Clarke's Orbit or Clarke's Belt in recognition of his foresight and vision.

MICROWAVE OVENS

Percy Lebaron Spencer (1894–1970) was only eighteen months old when his father died and his mother soon shipped him off to live with her sister and brother-in-law. By the time he was seven years old his uncle had also died, leaving the young master Percy alone with his aunt. When he reached twelve years of age Percy had to leave school and find work at the local mill, where he put in daily shifts, between sunrise and sunset, in order to support himself and his aunt. At sixteen Percy heard of another local paper mill that intended to modernize to the extent of installing electricity and, as nobody in his remote community could tell him anything at all about electricity, Percy began to read as much as he could find about the new, emerging technology. By the time he

40

applied for a job at the mill he had learned so much that he found himself becoming one of only three people who were engaged to install the new power supply, despite having never received any training or even formally studied the theory of electricity. Percy, of course, had never even finished his schooling.

By the time he had turned eighteen he had picked up a newspaper one morning to read about the disaster of the *Titanic*'s maiden voyage. Whilst most people were talking of the sinking of the indestructible ship, it was the work of the wireless communications operators that caught Percy's attention and he immediately decided to enrol in the US Navy so he could learn all about the new, fascinating wireless communications technology. As soon as he enlisted he learned all he could about radio and later recalled, 'I just got hold of a load of textbooks and taught myself whilst I was standing on watch at night.'

By the beginning of the Second World War, Spencer had become one of the world's leading radar tube designers and was working at Raytheon, a contractor for the US Department of Defense, where he was the head of the power-tube division. He was heavily involved in the development of the combat radar equipment that would become the Allied military's second-highest priority during the war, after the Manhattan Project (the creation of the atomic bomb). By this time Spencer was central to the development of microwave radio signals and was testing more efficient ways of manufacturing the equipment. One day, whilst standing in front of an active radar unit, Spencer noticed how the chocolate bar in his lab coat pocket became soft and then melted. It was not the first time this had happened, although the naturally inquisitive Spencer became the first to investigate the reasons behind such an event, especially as he himself detected no increase in the temperature around him.

Firstly he placed a plate of popcorn kernels in front of the radar set and was amazed to see them explode all around him within a minute. (Incidentally, popcorn would later become the world's favourite microwaved food.) He then experimented with other foods, including an egg that was placed in a tea kettle. It must have been an extraordinary moment when the egg suddenly exploded into the face of a colleague who was peering in to take a closer look. Percy Spencer then built the world's first microwave oven by placing a high-density electro-magnetic field generator into an enclosed metal box, which enabled safer and more controlled experiments, and his team observed the effects it had on various foods whilst monitoring the temperatures and cooking times. He realized the consequences of his accidental discovery and Raytheon filed for a patent on 8 October 1945 for a 'microwave cooking oven' that would be called a Radarange.

The first Radarange ovens were five-and-a-half feet high and weighed over half a ton, which meant they could only be used in places where large amounts of food needed to be cooked in a short space of time. Worse still, they cost nearly $3,000 and were only bought by railroad com-panies, ship builders and large, low-budget restaurants. The most successful of these was the Speedy Weenie hot-dog vending machine that was installed at New York Central Station in January 1947 and dispensed 'sizzling delicious' hot dogs in seconds flat. However, the food reviewers were quick to point out that French fries were never crispy and any cooked meat failed to brown. In some cases it even appeared to be uncooked. But the biggest blow to the microwave oven came when the Chairman of Raytheon, Charles Adams, was told by his personal chef that he would resign if Adams insisted on his food being cooked in the Radarange. Slammed by food lovers and chefs alike, the microwave oven looked to be a certain failure and it would be another twenty years before, in 1967, a home-kitchen version became available for only $495. Even so, it would be a further fifteen years before Percy Spencer's accidental invention became generally affordable and an indispensable part of everybody's kitchen. Today over 90 per cent of homes in the Western world have at least one microwave oven in their kitchens thanks to one single and very important chocolate bar.

That the automobile has practically reached the limit of its development is suggested by the fact that during the past year no improvements of a radical nature have been introduced.
Scientific American magazine, 2 January 1909

THE FIREMAN'S SAFETY HOOD
The threat of fire has been a problem for town planners and developers for centuries. The Emperor Nero recognized it and far from fiddling as Rome burned, in AD 64, he insisted on a new city development plan that would make it easier for fires to be fought if one ever broke out in his city again. The Great Fire of London in 1666 was another example of where firefighters could not get close enough to the flames to extinguish them and had to pull buildings down as fire-breaks instead. The problem, of course, is that men could never get close enough to a burning building without losing their vital air supply or

being overcome by smoke and other fumes. Attempts to make the work of firemen safer over the years included protective clothing, dome-shaped helmets and leather boots, but none of these helped a fireman to breathe when close to a fire, let alone inside a burning building. This was especially true of mine fires where there were usually fatalities.

Early, rudimentary regulations included instructions for firemen to grow their whiskers long and then soak them in water before tackling fires. Stuffing wet whiskers into their mouths was supposed to help them breathe in smoke-filled buildings. In 1825 the Italian inventor Giovanni Aldini (1762–1834) designed a mask to provide both heat protection and fresh air and this led to new attempts in safety-equipment design. A miner called John Roberts produced a filter mask that was used throughout Europe and America and attempts were then made to attach air hoses to hand-pumped bellows. The development of galvanized rubber during the 1850s (see 'Vulcanized Rubber: Charles Goodyear') led to improved ideas and in 1861 James Braidwood (1800–61), creator of the modern municipal fire service, designed a self-contained air supply by attaching two rubber-lined canvas bags together that could be worn on a fireman's back and used in an emergency. However, these were cumbersome and required heavy corks to be pulled out and tubes fitted when needed. Hardly ideal in any emergency. Goggles, a leather hood and a whistle were also added to the standard equipment, none of which would save many lives in a serious fire.

In 1907 African-American Garrett Augustus Morgan (1877–1963), the son of former slaves, opened his own sewing-machine and shoe-repair shop in Cleveland, Ohio, and soon earned a reputation as a skilful engineer with a creative mind. He invented a belt fastener for sewing machines and, in 1908, founded the Cleveland Association of Colored Men. The following year he opened a ladies clothing store and employed thirty-two people to manufacture his popular designs. One morning, in 1910, Morgan read about the Great Fire, also known as 'The Big Burn', which destroyed nearly three million acres of woodland in Washington, Montana and Idaho, during which there were eighty-seven fatalities; seventy-eight of them were men sent to fight the fire and who died of smoke inhalation.

After reading the reports Morgan sat down and tried to find a way of making firefighting safer, focusing on the supply of fresh air for those closest to danger. Within two years Morgan had invented a safety hood and applied for a patent in 1912, but none of the official government departments were interested in his design. Despite forming the National Safety Device Company in 1914, Morgan was forced to concentrate on his grooming products, which included hair-straightening cream, hair dye and afro combs. But he remained confident in his safety-hood device, which used a wet sponge to filter and cool fire fumes, and had a portable hose that dangled at a firefighter's feet, sucking in the layer of clean air that gathered nearest the floor of a smoke-filled building. He achieved moderate success with his device by using white actors to demonstrate it, whom he allowed to earn the credit for it around the country. Sometimes, however, he would thrill an audience by dressing as a full-blooded Indian (called Big Chief Mason) and entering buildings filled with noxious fire gases or tents full of manure. Big Chief Mason would amaze his audience by remaining inside for up to twenty minutes before emerging unaffected.

It would be a tragic event in 1916 that finally earned Morgan's safety hoods the international acclaim they deserved. In that year an explosion under Lake Erie (see 'Clinton's Ditch') trapped many men, and several attempts to reach them failed as rescue workers themselves were overcome by fumes and then became victims. Nobody else wanted to enter the tunnel, although one member of the rescue team remembered seeing a demonstration by Big Chief Mason and sent a messenger, during the night, to Garrett Morgan to persuade him to bring as many of his masks as he could carry. Morgan reacted so quickly that he arrived on the scene, with his brother Frank, still wearing his pyjamas. But they did have four of the safety hoods with them. However, the majority of the rescue team were sceptical of Garrett's invention and, given that several other members of their team had failed to return from the shaft, most of them refused to venture inside, safety hood or no safety hood.

The Morgan brothers, on the other hand, had no such reservations and the pair of them immediately pulled the hoods on themselves and disappeared down the shaft with the only two volunteers they could muster. The minutes passed and the tension rose until Garrett and Frank emerged with two members of the previous rescue effort over their shoulders. Two more soon followed and, as confidence began to grow, other men donned the safety hoods and raced inside. More survivors emerged and, later, the bodies of those who didn't were recovered. Garrett Morgan himself made four trips into the tunnel and many lives were saved.

Sadly, the Cleveland city officials and the newspapers failed to recognize the Morgan brothers' bravery and also managed to avoid mentioning that Morgan was responsible for the equipment that had been so successfully used. Officials appealed to the Carnegie Hero Fund Commission to present medals to many of the men involved in the rescue but excluded the Morgan brothers in what was generally regarded as a racially motivated snub. Happily, however, other rescue workers and a group of Cleveland citizens did acknowledge the brothers and presented them with diamond-studded gold medals in 1917. Morgan was also awarded a medal by the International Association of Fire Engineers and made an honorary member in recognition of his invention, which was then modified and made part of the standard equipment for a firefighter. Various improved versions have been part of the fireman's kit ever since.

The fire safety hood would not be Garrett's last contribution to modern society. As the automobile began competing for highway space with bicycles, pedestrians, horse-drawn carts, coaches and herded animals, Garrett was alarmed to personally witness a serious accident at a major road junction. He immediately set about a solution to the growing traffic free-for-all and began experimenting with signalling devices in 1913. It would be in 1922 that the inventor patented his first mechanical signalling system that could easily be operated by one man and a crank handle. With that, the modern-day traffic light had been invented. In his later years, suffering ill-health and nearly blind, Garrett Morgan continued to experiment and work on new inventions, including the self-extinguishing cigarette, in an attempt to further reduce the risk of fire throughout rural communities that still relied on wooden buildings. His ingenious idea employed a small plastic pellet that was filled with water and placed at the filter. Any discarded cigarette, Morgan pointed out, would self-extinguish as soon as the pellet burned through.

Despite little recognition during his own lifetime Garrett Morgan was later honoured by his home city in the shape of the Garrett A. Morgan Cleveland School of Science and the Garrett A. Morgan Treatment Plant. There is also an elementary school of the same name in Chicago, Illinois. There are numerous streets named after him throughout America, and in 2002 Garrett A. Morgan was included in a list of *100 Greatest African Americans*. And quite rightly too.

How Wrong Can You Be?
In January 1970 *Life* magazine published a feature article that predicted growing air pollution would reduce the amount of sunlight reaching the surface of the Earth by at least one half. Whilst noting that some people may disagree, the feature argued that 'scientists have solid experimental and historical evidence to support the prediction'.

THE PARACHUTE
The earliest evidence of a parachute design can be traced back to the Renaissance and an anonymous Italian manuscript that has been dated to 1470. Which is strange since buildings were not particularly high at that time and, as the first documented balloon flight was not made until 8 August 1709, there was nothing really to jump from that

would require a device to slow down a descent, apart from the odd cliff. Even so, Leonardo da Vinci (1452–1519) sketched a design for a parachute in his notebook in 1514 and a century later a Croatian Catholic priest by the name of Fausto Veranzio (*c.*1551–1617) borrowed details from da Vinci's design to build a rigid-framed parachute, which he used to jump from St Mark's Campanile in Venice during 1617. Veranzio's written description of his invention, and accompanying sketches, reveal his device to be called 'Homo Volans' (the Flying Man.) Veranzio is also credited with inventing the first metal-arched bridge and known for improving windmill designs. Although very little is recorded about his death or his attempt to jump safely from the Bell Tower of St Mark's, we do know that both events happened in the same year so we could make an educated guess about what happened.

Paolo Guidotti had made an earlier attempt, in 1590, to adapt da Vinci's design but only succeeded in falling from the roof of his house and breaking his leg. Veranzio's achievements were eventually recorded in 1648 when they were included in the *Mathematical Magick or, the Wonders that may be performed by Mechanical Geometry* written by John Wilkins, the Secretary of the Royal Society in London.

It would be a very long time until another attempt was made to float safely to earth from a great height. One hundred and sixty-six years to be exact when, on 26 December 1783, Frenchman Louis-Sébastien Lenormand (1757–1837) made what is widely regarded as the first witnessed and controlled descent. He jumped from the tower of the Montpellier Observatory while hanging on to a fifteen-foot rigid-frame 'umbrella design'. (He had actually practised by jumping from the top of a tree holding two umbrellas.) Lenormand was, in fact, demonstrating his design as a means for people to escape tall buildings in the event of a fire, and it was he who coined the term 'parachute' by connecting the Greek word '*para*', meaning 'against', and the French word '*chute*', which translates as 'fall'. Even though the demonstration was successful, and witnessed by Joseph-Michel Montgolfier (1740–1810) who invented the hot-air balloon, nobody else was prepared to give it a try. Except for Jean-Pierre Blanchard (1753–1809), another Frenchman, who dropped his dog in a parachute basket from a balloon in 1785. But that doesn't count. In 1793 Blanchard also claimed to have escaped from a burning hot-air balloon by parachute but nobody saw it and nobody believed him. And, I imagine, as the French Revolution was the main news of the time, nobody much cared either.

> The horse is here to stay, but the automobile is only a novelty, just a fad.
>> President of the Michigan Savings Bank, warning Henry Ford's lawyer not to invest in the Ford Motor Company

However, Jean-Pierre Blanchard is credited with designing the first frameless, and foldable, parachute made of silk during the late 1790s, but there is no record of him ever using it himself, only his dog. Instead, that dubious honour fell to yet another Frenchman, André-Jacques Garnerin (1769–1823), who became the first person to jump without a rigid frame when he dropped from a hot-air balloon at 3,000 feet on 22 October 1797.

Although he was safely seated in a wicker basket, attached to the silk parachute, his design swung violently and out of control in midair before landing heavily in a field. Happily, he did emerge unscathed. His wife Jeanne was suitably impressed and had a go herself a few years later. As word spread around Europe, the Garnerins were invited to perform a series of demonstrations which inspired the popular English ballad: 'Bold Garnerin went up / Which increased his Repute / And came safe to earth / In his Grand Parachute.' (No, I have never heard of it either.) Although their efforts did inspire others, it wasn't very many others and the idea of parachuting as a safety option from either a building or balloon remained unpopular.

In 1837 Robert Cocking (1776–1837), an English artist and amateur scientist who had witnessed Garnerin's demonstration jump in London during his tour of 1802, decided to have a go himself. Despite being sixty-one years old and having no experience, he designed his own parachute and persuaded the organizers of the Grand Day Fete at Vauxhall Gardens in London to promote his debut effort as their main attraction. At approximately 7.35 p.m. on 24 July 1837, Cocking rose high into the air, dangling in a basket below a hot-air balloon. He had intended to release his invention from the balloon at a height of 8,000 feet but a miscalculation meant he had only reached 5,000 feet by the time he found himself floating over Greenwich, several miles away. Realizing he was running out of daylight and facing the prospect of aborting the attempt, Cocking released his chute. It was immediately obvious he had made further miscalculations and the date of his death will confirm the outcome of this attempt. Further efforts were made during the nineteenth century, with similar results.

In 1912 an Austrian-born French tailor called Franz Reichelt (1879–1912), now sometimes referred to as the Flying Tailor, famously designed a flying-suit parachute in the hope that early aviators would buy them to save their lives if they were ever forced to jump from their aircraft at height. Which, in those early days, many of them were. In order to attract maximum publicity for his design Reichelt persuaded the Paris city authorities to allow him to demonstrate his invention by dropping one of his tailor's dummies, wearing the flying suit, from the Eiffel Tower. However, after being granted permission he arrived at the tower on 4 February at 7 a.m. and announced to the gathering crowd that he intended to jump himself. His friends and some of the spectators tried to dissuade him but Reichelt was adamant and, in front of a huge crowd, including a few early news cinecameras, he joined the long list of Frenchmen who had leapt into the unknown. Once again, the date of his attempt and the date of his death will reveal the outcome of this particular demonstration.

Nearly 300 years after Veranzio's first attempt from the Venetian Bell Tower people were still saying, 'This isn't a very good idea, is it?' But other people were still trying to perfect the parachute. In fact, during 1911 and 1912 there was a flurry of activity and, apparently, no shortage of people prepared to throw themselves out of balloons or airplanes in an attempt to demonstrate their design. The race was on.

One of these people was a chap called Štefan Banič (1870–1941), a Slovakian immigrant, born in Austria-Hungary, who was working as a coal miner in Greenville,

Pennsylvania. Banič had also spent time at evening classes learning engineering in an effort to free himself from the manual labour he was facing for the rest of his life. One day in 1912, whilst walking home after his shift at the mine, Banič was terrified to see a plane fall from the sky and crash right in front of him. The pilot stood no chance of survival. The event haunted Banič and he began to consider ways in which the pilot could have been saved. It had only been nine years since the Wright brothers had taken their first flight and the aviation industry was still in its infancy. Banič thought it was the future, if only it could be made safer and pilots stood a chance of survival should their motor break down in midair, which was more common in those days than it is imaginable in modern times. The idea for a parachute was not a new one, although none of them had worked reliably. At least, none that could be worn by a pilot in the cockpit of an aircraft.

Štefan Banič was convinced that the early designs could be improved upon and he began to hand-stitch parachute devices in a barn. He used canvas for the canopy and attached expanding poles that could either extend or contract, and he also designed a harness that could be worn around the waist. Banič thought his parachute would be safer and more reliable than previous efforts if it could be steered, or even held open, as it could be if it was worn below his arms. Neighbours became used to the man jumping off the roof beams in a barn wearing an odd-looking skirt but, over time, Banič began to refine his ideas. Eventually, he had what he thought was a work-able prototype but he was unable to apply for a patent without first demonstrating how his parachute actually worked. And he had to prove it was effective from a much greater height than the roof of his barn. So he planned a very public and equally risky demonstration by leaping from the top of a Washington building, located opposite the US Patent Office. On 3 June 1914 the crowds gathered, including military observers and Patent Office officials, as Banič climbed to the top of the fifteen-storey building, balanced on the edge of the roof and, after taking a deep breath, jumped.

To everybody's astonishment his descent was perfect and Banič landed before them in a safe and controlled manner. His publicity stunt was a complete success and the crowds were thrilled; his patent was awarded (no. 1,108,484 on 25 August 1914) and the military were immediately interested. It was the beginning of the First World War and the idea of a device that could get a pilot out of a broken aircraft and safely back to the ground was high on the list of priorities. So Štefan Banič immediately donated his patent to the American Society for the Promotion of Aviation and the newly formed Army Signal Corp, who developed his design further, making it more practical for airmen. The parachute soon became standard issue for all Allied pilots and countless lives were saved as a result. The First World War was a time when many entrepreneurs and inventors were cashing in (see 'It Was a Close Shave') but Banič received neither recognition nor reward for his vision, despite his invention becoming one of the most important in the history of warfare, if not aviation itself.

Today, Banič's parachute skirt is nothing more than a museum curiosity but it did start a revolution in aviation safety. After the war Banič returned to his homeland, which was by then called Czechoslovakia, where he remained until his death in 1941, just as the first

54

parachute regiments were being formed in the early days of the Second World War. What began as a simple safety device had, by then, developed into a way of delivering deadly foot soldiers into otherwise unreachable areas. In 1970 a memorial was erected at Bratislava Airport in Slovakia to the man who finally demonstrated that the parachute, first considered exactly 500 years earlier, was indeed a good idea after all.

Automobiles will start to decline almost as soon as the last shot is fired in World War II. The name of Igor Sikorsky will become as well-known as Henry Ford's, for his helicopter will all but replace the horseless carriage as the new means of popular transportation. Instead of a car in every garage, there will be a helicopter. These 'copters' will be so safe and will cost so little to produce that small models will be made for teenagers. These tiny 'copters, when school lets out, will then fill the sky as the bicycles of our youth filled the pre-war roads.

1943 prediction from respected aviation journalist
Harry Bruno (1896–1978)

ANCIENT INVENTIONS WE STILL USE EVERYDAY

Scissors were first invented around 1500 BC and early examples have been discovered in ancient Egyptian ruins. To begin with they were forged out of a single piece of metal with a pair of blades attached to a U-shaped, spring-like handle. It would be another 1,600 years before the Romans developed a more practical design by using two cross-blades that were attached in the centre by a screw or rivet, in exactly the same way as they are still produced. They were successfully used by tailors and barbers although, despite the Romans introducing their technology throughout Europe until the sixth-century AD, they did not come in to common use for another thousand years, when they were finally produced by European countries during the sixteenth century.

Eye Glasses were first recorded by the Romans, although the earliest designs to be discovered were found in China and only date from the thirteenth century. At the same time two Italians were working on lenses that would correct far-sighted vision. Salvino D'Armate of Pisa and Alessandro Spina of Florence were both producing early lenses that perched on the bridge of the nose. Lenses that corrected near-sightedness, or myopia, would be developed a century later, during the 1400s. It would be two hundred years before somebody had the idea of connecting the lenses to arms that hooked behind the ears, making them far more practical for users. However, progress remained slow until the American politician and inventor Benjamin Franklin worked out a way to combine convex and concave lenses to correct both near- and far-sightedness (bifocals) in 1775, and they became the blueprint for all subsequent designs for eyeglasses. It was a 1,700-year development period for the spectacle before they finally established themselves as an indispensable part of modern life.

The **Compass** was first produced in the Chinese Han Dynasty at some time between the second century BC and the second century AD. The Chinese first noticed that lodestone, a naturally magnetic variety of ore, always pointed in the same direction if it was suspended in water, often on a leaf or something else light enough to float. But it is doubtful the Chinese realized this direction was magnetic north as little was known about global navigation at the time. And their device, it seems, was never even used for navigational purposes. Instead, it was used to determine the precise direction of buildings and roads in a sort of early demonstration of feng shui.

Gunpowder was first produced in China when at least one brave soul discovered that a mixture of saltpeter (potassium nitrate), powdered charcoal and sulphur had explosive qualities. One can only imagine what ran through the mind of the man who first realized this. Hopefully it wasn't his sandals. Initially the Chinese used their invention to make signal flares that could be seen over long distances. Later they developed the modern fireworks and, eventually, one bright spark realized they could use them to launch arrows. These fire rockets were made by filling bamboo tubes with a mixture of gunpowder and iron shards (shrapnel). An arrow was then attached, a fuse was lit and the whole rocket then fired from a bow. The invading Mongols must have had quite a shock when the first of these began landing among them as they charged the Chinese positions during their failed early invasion attempts.

ARCHIMEDES (287–212 BC)
Legend has it that the Greek mathematician and engineer Archimedes of Syracuse was asked by King Hiero II whether the gold he had been sold to decorate his new temple had been mixed with a cheaper metal by a dishonest gold merchant. The King had given Archimedes his crown, which he knew to be of solid gold, and commissioned him to test the two samples. This left Archimedes with a problem as he dared not chip or damage the crown in any way and melting it down was clearly

out of the question. Instead, he had to come up with a way of calculating its density in order to compare it with the merchant's sample. As he pondered the problem Archimedes took a bath and noticed how the water level rose as he sat in the water. Famously, and as every student knows, he realized that this was a way to determine the density of the crown as it would displace a volume of water equal to the volume of the crown. This was when, as the story goes, he jumped out of the bath and ran naked down the road shouting 'Eureka', which is Greek for 'I have found it'. (If you didn't already know that then you should have been paying more attention in class.)

Indeed, he had found it: he had invented a method of measurement known as Archimedes's Principle, which is that any object wholly or partially submerged in fluid is buoyed up by a force equal to the weight of the fluid displaced by the object. In other words, it is how those great big ships stay afloat. And so, using that principle, Archimedes proved the gold had been mixed with a cheaper, and less dense, silver. It isn't known what the King did to the merchant, but we can guess he didn't take his receipt back for a refund. What we do know is that Archimedes did design the *Syracusia*, King Hiero's custom-built super-ship, the grandest of its day, and in doing so invented a method of moving water upwards in order to pump out the bilge. That screw-shaped blade, housed inside a cylinder, became known as the Archimedean Screw and revolutionized design and architecture from that moment onwards. After all, how do you think the Romans had all those fountains and aqueducts without a means of moving water upwards and against the pull of gravity?

In modern times the Archimedean Screw is still used

for pumping liquids and shifting grain or coal dust; it is the influence for the screw propeller which, over the years, has transformed ocean-going vessels. The first steamship that used a screw propeller, launched in 1839, was called the SS *Archimedes* in honour of the man who made it all possible. Archimedes also invented the claw, which was a crane-like device used to drop onto an enemy ship and lift it out of the water, a forerunner of today's modern mechanical crane. He is also credited with developing the heat ray, which was a bank of mirrors facing the sun to direct its rays onto enemy ships approaching the Syracuse harbour. They burst into flames within seconds. He also invented the block-and-tackle pulley system and the catapult, which became an effective weapon of mass destruction for many centuries.

HOISTED BY THEIR OWN PETARD: INVENTORS KILLED BY THEIR CREATIONS

The next section is full of short stories of inventors who were killed by their creations. Some of these ideas were immediately abandoned and it is easy to see why. Others, however, are still in everyday use. The parachute is not mentioned because it is covered earlier and, more importantly, too many of the inventors sacrificed themselves to do them justice here.

The Cure for Cancer

Marie Curie (1867–1934) is quite possibly the most famous woman in the history of science. During her lifetime she received numerous awards, including the Nobel Prize for Physics (1903) and the same honour for Chemistry in 1911, the first scientist (male or female) to ever achieve such an honour. Curie went on to receive many more awards, even after her death. Born in Warsaw, Poland, Marie Skłodowska studied at the Sorbonne in Paris where her lack of funding, and diet of bread and tea, led to regular health problems. However, she did manage to complete her master's degree in physics in 1893 and earned a commission to study various metals and identify their magnetic properties. And this is where her fortune changed because, in need of a place to work, a friend introduced the young Polak to French physicist Pierre Curie, a scientist with a growing reputation and, more importantly, access to a laboratory. The pair were soon romantically linked and formed a formidable scientific partnership.

Before long, Marie Curie was experimenting with uranium rays after studying the work of Henri Becquerel and Wilhelm Röntgen (see 'The X-ray Is a Hoax'), and coined the word 'radioactivity' in the process. Her new husband soon suspended his own research to join his wife's experiments. By 1898 the couple had identified a new radioactive element that they called Polonium, in honour of Marie's homeland. In 1903 Pierre's life was cut tragically short when he stepped in front of a horse-drawn carriage in Paris and suffered a fatal fractured skull. Heartbroken, Marie continued obsessively with her research. During the First World War she actively promoted the use of portable X-ray machines, which became affectionately known as 'Little Curies', and she was well known for carrying test tubes full of uranium around in her lab coat pocket. This habit, and her devotion to science, would ultimately lead to tragedy when, in 1934, she was admitted to the Sancellemoz sanatorium in Passy, France, where she died on 4 July of aplastic anaemia (leukaemia) caused by prolonged exposure to radiation.

It was Marie Curie's ground-breaking work that led to the successful use of radiotherapy in the treatment of cancer during the twentieth century.

Advanced Gunships

Cowper Phipps Coles (1819–70) was both a British naval captain and inventor. At the age of only eleven years, in 1830, he joined the Royal Navy and served with distinction throughout many military campaigns, not least at the Siege of Sevastopol of 1855 during the Crimean War against Russia. By the following year Coles had been promoted to commander of the paddle sloop HMS *Stromboli*, which patrolled the Black Sea until the end of the conflict. During this time Coles and a few other officers designed and built a 15-metre raft out of empty barrels, which they christened *Lady Nancy*, and lashed a thirty-two pounder cannon to her deck. With this vessel Commander Coles was able to manoeuvre the gun into shallow waters and attack the Russian supply stores at Taganrog, a task previously impossible with their own deep-water ships. The action made a hero of Coles and he began to experiment with other designs for military technology, which impressed the Admiral of the Fleet enough to send him to London and explain his ideas to the powers that be in Greenwich. Coles's inventive plans were well received, although the war ended before any of them could be put into use.

Undeterred, Cowper Phipps Coles, by then a full captain, began designing turret ships, which included a revolving platform that guns could be set upon and then quickly aimed in any direction, a marked improvement on the previous ships of the day that had to slowly turn on the wind until their guns faced the target. Coles's rotating gun platforms were definitely the future of gunships and his patent was filed on 10 March 1859. His designs received enthusiastic support, not least from Prince Albert who encouraged the Admiralty to build the first turret ship in 1862. However, Coles was only permitted to design the turret; the ship upon which it sat was left to the chief Navy engineer, Isaac Watts. Coles then improved his design and proposed other ships that he was eager to build himself but he was repeatedly denied the opportunity.

Eventually, and after growing public support, Coles was allowed to supervise the building of his latest design, although the Admiralty Board was divided in their decision. In fact, the Board was so divided that the project was not officially 'approved', but stamped 'not objected to' instead. Construction of the HMS *Captain* began in 1867 and, despite the official concerns of the Navy's new Chief Constructor, Edward Reed, over her seaworthiness, successful trials were completed in January 1870. Reed, however, remained unconvinced and eventually resigned over the dispute in July of the same year. Coles, on the other hand, was vindicated when his new ship was officially launched the following month with its inventor on board. However, another month later, as she sailed in the open sea, the weather deteriorated, a gale blew up and the hurricane decks, above the turrets, caught the wind. The HMS *Captain* capsized as she turned about. Coles and nearly 500 English seamen went down with her and were lost to the sea.

The Original Spaceman

Wan Hu was a Chinese official of the Ming Dynasty during the sixteenth century who has passed into legend for attempting to become the first man to reach outer space. Hu's ambitious plan was to employ China's advanced, gunpowder-based firework technology to build rockets that would lift a chair into the above and beyond with apparent ease. What his plans were for returning safely to earth remain unknown. When his testing was complete and his big day arrived, Wan Hu had forty-seven of the most powerful rockets attached to a grand chair and, wearing his most regal outfit, the inventor climbed aboard and prepared for lift-off.

Forty-seven faithful servants then lit each of the fuses and scurried for cover. Peering from behind a wall they later reported a huge explosion and, once the smoke had cleared, no trace of the fearless inventor, or his chair, could be found. Nothing was ever heard of him again and, in his honour, one of the craters on the dark side of the moon is named the Wan Hoo Crater. Some faithfully believed that it was his actual landing spot. Others, on the other hand, are perhaps more realistic in their conclusion.

It's All in the Blood

Alexander Bogdanov (1873–1928) was a controversial figure and an old Bolshevik who had resigned from the party long before the Russian Revolution, but he had made a comrade of Vladimir Lenin, who continued to respect him as a scientist. He also valued his opinion. He had been a rival to Lenin during the first decade of the twentieth century but eventually came to support him. Lenin, for his part, agreed to forgive Bogdanov as long as he stayed out of politics in the future and concentrated on science instead. Bogdanov then spent the next twenty-five years working as a physician, economist, scientist, teacher and inventor (among other things), and early enthusiast of cybernetics. He also became the founder of the world's first institution that devoted itself entirely to the research of what we now call blood transfusion. For the first decade, after leaving the Bolsheviks, Bogdanov was constantly on the move. A marked man, he was pursued by supporters of the Russian Tsar and had to adopt a variety of pseudonyms before returning to his original name after the storming of the Winter Palace.

In 1924 Bogdanov began experimenting with blood transfusions on himself in an attempt to stay young forever or, at least, achieve some sort of longevity. Lenin's sister Maria Ulyanova was one of those who volunteered to take part in these experiments and Bogdanov himself, after eleven successful transfusions, recorded that his eyesight had improved and his hair loss had decreased. Other volunteers noted that Bogdanov appeared to be at least ten years younger after the transfusions. In 1925 he founded the Institute for Haematology and Blood Transfusions and continued his good work. However, far from increasing his lifespan, the science of blood transfusion was soon to cut it short as he took the blood for his twelfth transfusion from a student who was unknowingly carrying tuberculosis and malaria. Bogdanov was soon infected and overcome with illness, which killed him on 7 April 1928. Ironically, the infected student was given Bogdanov's own blood as a replacement and he later made a full recovery.

The Flying Taxi

Michael Dacre (1956–2009) was a man of vision. He understood the needs of modern society, especially in built-up urban areas, and in 1998 he invented a flying taxi known as a Jetpod, which could take off and land in very short distances. He thought that with top speeds of over 350 mph it would be perfect for executive transport, which currently relied on slower, and noisy, helicopters. Dacre claimed that his Jetpod would take passengers from Heathrow to the centre of London in under four minutes at a cost of only £50 and spoke of his park-and-fly concept with great enthusiasm: 'Jetpods are meant to be a workhorse. A taxi cab in the air for on-demand free-roaming traffic.' He went on to explain they would have over-wings that would reduce noise, and Jetpods would be any city's cheapest and fastest form of travel.

Dacre was clearly a latter-day aviation pioneer and his invention drew an enthusiastic response. However, the plan failed to take off, so to speak. On 16 August 2009 the 53-year-old inventor was preparing for his first test flight at the remote village of Taiping, 150 miles north of the Malaysian capital, Kuala Lumpur. A local villager, who was collecting shrimps from a nearby pond, later described to the aviation authorities what he had witnessed: 'I had seen it earlier going down the runway three times, but it couldn't take off. Then, on the fourth run, it got about two-hundred metres into the air and then shot vertically into the sky before veering off to the left and falling to the ground.' The plane exploded into a ball of flames but the firemen, who were very quickly on the scene, managed to extinguish them. Sadly, Michael Dacre had died on impact, killed by the invention he had hoped would change city travel.

Underwater Travel

Horace Lawson Hunley (1823–63) was born in Tennessee and raised in New Orleans, a good old southern boy. He studied law and then became a member of the Louisiana

State legislature before the American Civil War broke out, and Hunley turned his attention to the invention of weapons that he hoped would help the Confederate Army prevail. And in particular, he devoted himself to the technology of submarines. Working with designers Baxter Watson and James R. McClintock, Hunley completed his first effort, called the *Pioneer*. However, New Orleans was soon to fall to the wrong army and the *Pioneer* was scuttled to prevent its capture and its design features noted. His second effort was built in Alabama but she sank on her maiden test run in Mobile Bay. But Hunley was undeterred and, convinced his invention would be of major advantage to Confederate seamen, he started again with a third design.

This time he was forced to finance the project himself and his manually propelled design was successfully tested in the same Mobile Bay as his previous effort. Named in his honour, the *H. L. Hunley* was then secretly transported to Charleston Harbor in 1863 where it was commissioned to protect the town from an attack by sea. On 15 October 1863 the *Hunley*'s skipper, Lieutenant Dixon, was on leave and so her inventor decided to carry out a series of tests himself. And whilst diving into deeper water, in an attempt to navigate her beneath an anchored ship, Horace Hunley managed to bury her nose into deep mud at the bottom of the harbour. Seven crew members drowned and Hunley himself, who was at one of the escape hatches, ended up suffocating before he could be rescued. It eventually took divers three days to reach the stricken *Hunley* and remove the bodies.

The Life-saver that Killed its Inventor

Henry Winstanley (1644–1703) was born in Saffron

Walden, Essex, the son of a gamekeeper who would later take a position at Audley End House, the seat of the Earl of Suffolk. The younger Winstanley also worked at Audley End House, first as a porter and later as secretary as the home became a royal palace for the use of King Charles II whilst he was idling away his time at Newmarket, attending the horse racing. Winstanley also took a keen interest in the practice of engraving, and during his Grand Tour of Europe (1669–74) he developed an enthusiasm for continental architecture. On his return he spent ten years creating a detailed set of architectural engravings that survive today as an important historic record of English manor-house design. He also designed a set of playing cards that proved to be very popular and sold handsomely. In 1679 he was appointed the Clerk of Works at Audley End House and became known, throughout society, for his ingenious mechanical and hydraulic designs and for his love of gadgets. Indeed, he filled his own house at Littlebury with fascinating new designs so that it became known as the Essex House of Wonders, a popular tourist landmark for well-to-do and curious travellers.

In 1690 Winstanley opened his Waterworks Theatre in London's Piccadilly, which proved to be instantly popular and produced a great deal of profit, much of which its owner invested in ships. England, at the time, was becoming one of the greatest seafaring nations on earth and owning galleons was considered the shrewd investment of its day. Ultimately, Winstanley owned five although they were not without their risk. One risk in particular, the Eddystone Reef, a series of rocks fourteen miles off the coast of Cornwall, had been busily claiming hundreds of trading ships at the cost of thousands of lives. Two of these ships were part of Winstanley's own fleet and, after

67

the second loss, he approached the Admiralty with an audacious plan to build a structure directly onto the rocks and erect a warning light at the top. And he also persuaded the authorities in London that he was the man to build it.

In 1696 Winstanley's plans were approved and work began on 14 July of that year. He had, as it turned out, invented the modern lighthouse. Building it was not without its challenges as one morning, while the foundations were being completed, a French privateer ship destroyed the structure and kidnapped Winstanley, taking him off to France where he would be held to ransom. This episode did not last long for, as soon as French King Louis XIV heard what had happened, he ordered Winstanley's release and reminded his subjects; 'We are at war with England, not with humanity.' Winstanley returned to begin his project again, although, when it was completed, the designer was unimpressed with the wooden structure's resilience to the first severe gale. And so he ordered it down and started again.

At the second attempt Henry Winstanley completed a solid, part-stone structure with a fine continental design complete with a luxury stateroom and living accommodation. Winstanley proudly announced to the world that his lighthouse could withstand any of the natural elements and that his biggest wish was to be 'inside it during the greatest storm that ever was'. Over the following five years there were to be no ships lost to the Eddystone Reef and Winstanley's invention was repeated at numerous locations around Great Britain.

Sadly, Winstanley was soon to have his wish come true, for on 7 December 1703 one of the severest storms ever recorded descended upon the south of England. In

London over 2,000 chimneys fell down and the roof was blown off Westminster Abbey. At the time Winstanley was safely in his lighthouse carrying out a survey when the high seas cut him off from land. And so he hunkered down for the night, confident that his lighthouse would provide shelter. The following morning, as dawn broke, the Eddystone Lighthouse was gone, and so was Winstanley.

Strangled by His Own Bed

Thomas Midgley Jr (1889–1944) graduated from Cornell University in 1911 with a degree in Mechanical Engineering, and he began his working life as a designer for the National Cash Register Co. in Daytona, Ohio. Within a year his father had recruited him for the family firm, Midgley Tire and Rubber Co., where he settled into life as Chief Engineer, although it was to be a short-lived appointment as the company soon failed. So, in 1916, Midgley Jr was left, once again, to his own devices.

Later that same year he took a post in the research division of the newly formed Dayton Engineering Laboratories Co., a subsidiary of General Motors, with esteemed engineer Charles Kettering. Their collaboration proved to be so successful that Kettering later described 'Midge' as his greatest discovery. It was during this time that Midgley turned himself from mechanical engineer into chemist, thanks to his love of experimenting, and had soon identified tetraethyl lead, which reduced engine noise when added to petrol. He also discovered a method of extracting bromine from seawater, which prevented the lead from corroding engine parts. Midgley Jr also contributed greatly to the scientific research into natural and synthetic rubber and discovered how freon could be used as a non-toxic and non-flammable refrigerant. Freon would become central to the refrigeration and air-conditioning units that would be installed throughout the world.

Thomas later became vice-president of the Ethyl Corporation and a director of Ethyl-Dow Chemical Company. The American Chemical Society awarded him the Priestley Medal in 1941 for his achievements in the field of chemistry and he is, today, considered to be one of the most creative chemists of all time. However, his work had taken its toll: in 1923 he suffered lead poisoning as a direct result of his experiments with tetraethyl lead, and he spent a large part of that year convalescing in Florida. At the time, he wrote, 'I find my lungs have been affected and that it is necessary to drop all work and get a large supply of fresh air.' By the time he was honoured with the Priestley Medal he had already contracted a form of polio, which left him without the use of his legs. But he wasn't about to allow this to stifle his inventive mind. Nor would it temper his sense of humour for, in 1944, in an address entitled 'Accent on Youth', Midgley Jr pointed out that most of the great inventions had been made by scientists between the ages of twenty-five and forty. Two of his own great discoveries had been made at the ages of thirty-three and forty respectively. With that, Midgley urged the older scientists, like himself, to move over and allow their younger protégés to develop and realize their potential.

He concluded his speech with a short poem that thrilled the audience:

> When I feel old age approaching
> And my breath is growing short

My eyes are growing dimmer
And my hair is turning white
I lack the old ambitions
When I wander out at night
Though many men my senior
May remain when I am gone
I have no regrets to offer
Just because I'm passing on.
Let this epitaph be graven
On my tomb in simple style
This one did a lot of living
In a mighty little while.

Prophetic this may have been, for only a month later, at the age of fifty-five, Midgley was found in bed, strangled by the pulley and rope mechanism he had invented to help pull himself upright.

The Man who Invented the Modern Newspaper

William Bullock (1813–67) was born in Greenville, New York. He was orphaned shortly afterwards and then raised by his older brother. At the age of only eight he was apprenticed by his brother and put to work in an iron foundry as a machinist. Over the next decade or so his love of books, and fascination with mechanics, led to him running his own machine shop in Savannah, Georgia, by the time he was only twenty-one. During this time he invented a shingle-cutting machine, but he was unable to find a market for it and was bankrupted. However, his personal life was more successful as he married Angeline Kimble of Georgia who went on to bear him seven children. When Angeline died in 1850 he married her sister, Emily, who in turn produced six more mouths to feed.

In between wives Bullock worked productively and successfully invented cotton and hay presses, a seed planter, a lathe and a range of artificial legs. In 1849 his grain drill won him a prize from the Franklin Institute and his reputation was growing far and wide. During the early 1850s he took the position of editor at the *Banner of the Union* newspaper in Philadelphia and in 1853 he turned his attention to printing presses and built a hand-turned, self-feeding wooden press that he very quickly realized was a vast improvement on the hand-fed presses of the day.

In 1863 William Bullock applied for, and received, a patent for his new printing press, which he described at the time as, 'My improved machine for printing from moving type, or stereotype-plates, belongs to that class of power printing presses in which the paper is furnished to the machine in a continuous web or roll, and by which the sheets are severed from the web, printed on both sides and delivered from the machine "perfected". In my machine, however, there is but one delivery apparatus, which is simple in construction and works as rapidly as the machine can be driven and the sheets printed. So that by my invention the great obstacles to the rapid operation of the printing press are successfully overcome.' All this meant was that he had found a way of delivering paper into the press, from a great roll, that could be printed on both sides and then cut into sheets as it emerged from the other end. Quite an advancement over the man hand-feeding single sheets into a press, the technology that the printing industry had relied upon until then.

The press was self-adjusting, and it folded the paper and then cut it with precision. Bullock's printing press could produce up to 12,000 sheets per hour and later modifications increased that figure towards 30,000 per hour. It was a huge advantage for an industry that was growing quickly, with some newspapers boasting 100,000 daily circulations. By 1866, and the end of the American Civil War, Bullock was at the front of the daily newspaper industry and his printing presses were being installed, or copied, as fast as they could be produced.

Unfortunately, William Bullock would not live to reap the greater part of his rewards. On 3 April 1867, his company had installed a new press at the *Philadelphia Republic Ledger* newspaper. While setting it up Bullock crouched under the machine to kick a drive belt back onto the pulley. He missed and his leg was crushed as it was caught up in the belt mechanism. A few days later he developed gangrene and, on 12 April 1867, as he was undergoing amputation, he died in the operating theatre. It took a long time before William Bullock's inventiveness was formally recognized, although, in 1964, he was finally honoured as one of Philadelphia's geniuses and a bronze memorial was erected that reads: 'His invention of the rotary web press in 1863 made the modern newspaper possible.'

The First Man to Really Fly

Man's obsession with flight probably began as soon as he, or she, was able to walk upright. Although there is no evidence of prehistoric man jumping off cliffs in order to see what happened, it is safe to assume that one or two of them must have. The Greeks have a story (one you would remember if you had been paying attention in school) about Icarus and his father Daedalus, a master craftsman, who flew majestically through the sky with wings made of wood, bird feathers and wax. According to legend, Icarus flew too close to the sun, which then melted the wax, causing all the feathers to fall off, and he plummeted to his death. His less adventurous father, on the other hand, flew lower and landed safely. Apparently this all happened around 1400 BC and, for one reason or another, should not be taken too seriously. But there are cave drawings that archaeologists have dated as far back as 2500 BC depicting men with feathered wings, so who knows? We also have Robert Fabyan's sixteenth-century *Chronicle* that tells the tale of Bladud, King of the Britons, who made a set of wings and tried to fly from the top of the Temple of Apollo, in what is now London, and broke his neck. Again, we can't be certain but we do know he was mad as he not only believed he could communicate with the dead, but also that rolling around in pig-mud would cure leprosy. Another, more reliable, story tells the tale of a winged actor who tried to please Roman Emperor Nero in 60 AD by flying from a cliff-top straight into the ground. Marco Polo returned from China in 1295 AD with similar stories that had identical consequences.

But this doesn't seem to have stopped anybody from trying again. It certainly didn't stop the Prussian siblings Otto and Gustav Lilienthal, who spent their formative years studying the flight of birds, from strapping on wings and jumping off things. However, they had the good sense to make sure they were not too high to begin with by testing from the roof of their storage shed. In 1867 Otto, by then a professional engineer, had received patents for his mining designs and founded his own company dedicated to the manufacture of steam engines and boilers. But his obsession with birds and flight clearly continued as in 1889 he published his now famous book, *Bird Flight as the Basis of Aviation*. By then the 41-year-old inventor had achieved over 1,000 flights, and by 1891 he was regularly taking to the air with wings that were similar to a modern hang-glider. He even had his own 200-foot hill built near Berlin so that he could fly whenever he chose and in any direction, achieving flights of up to 250 metres. His work, drawings and, by then, even photographs of him in flight were widely circulated and influenced many of the world's contemporary designers. In America the Wright brothers closely studied the work and progress of Otto Lilienthal. He is widely regarded as being the first person to achieve what was then called 'manflight'.

Amazing engravings and photographs began appearing in magazines and newspapers all over the world and people marvelled at the sight of a human being with wings in mid-flight. Nobody had seen anything like it and even as late as 1895, Lord Kelvin, the first scientist to take a seat in the House of Lords, had declared 'heavier than air flight is not possible'. The following year Kelvin was quoted as saying, 'I have not the smallest molecule of faith in aerial navigation other than ballooning. I would not care to be a member of the Aeronautical Society.' Mind you, as previously mentioned, Kelvin also predicted 'no future' for the wireless radio. Even so, at that time he was generally thought to be right as most, if not all, aviation pioneers up until that point had died in the attempt (see 'The Parachute'). And yet there was Otto Lilienthal's picture gracing news-stands the world over, apparently defying scientific logic.

That was until 9 August 1896 when his glider dramatically collapsed and the pilot crashed 15 metres into the ground. He was immediately rushed to hospital in Berlin where he died the following day. His death at the hands of his own invention was widely regarded as a major setback in the progress of heavier-than-air flight, or man-flight, although it did not stop people from trying, as the Wright brothers themselves demonstrated only seven years later. Lilienthal's final words to his brother Gustav were: 'Sacrifices have to be made.'

FOOD

THE POTATO

Don't worry about silver and gold, the humble spud is far more valuable.

The young and ambitious Spanish Empire decided to send an expedition led by Hernán Cortés to sail across the Atlantic Ocean during the early sixteenth century, and their destruction of the Aztecs in 1520 provided unimaginable riches in terms of gold, silver and jewels for their royal paymasters. Subsequent successes changed the fortunes of the Spanish monarchy and altered the balance of power in Europe. As the news spread many more young Spanish conquistadors were being financed and encouraged to set sail for South America to see what they could discover for themselves.

By the time Francisco Pizarro González (1476–1541) landed in Peru in 1532, captured the Inca king and then offered his release for a 'room full of gold', the Spanish coffers were overflowing and other European countries like Holland were keen to follow suit. England and France, however, were slow out of the traps when it came to empire building as they were constantly at war with each other. Instead, the balance of power between the other European empires would shift continually as new discoveries were made and new wars broke out all over the globe. Gold and silver, it seemed, were the key to the future of Europe but, in actual fact, it was an almost over-looked discovery by the Pizarro expedition that would change the demographic of the entire world within a few centuries – that of the humble potato.

Prior to Pizarro's voyage the staple diet of most Mediterraneans consisted of wheat-flour, bread and beans, and although northern Europeans were used to root vegetables such as turnips, the idea that people would eat something that had been dug out of the ground revolted most Spanish adventurers of the age. However, Pizarro discovered that the principal energy source for the Incas and their predecessors were the tubers that could be frozen and kept in underground store-houses for long periods without decay. The *chuño*, as it was known, soon became the staple diet of the silver and gold miners across Peru who were providing vast wealth for the Spanish government. Returning sailors also carried *chuño* on board their ships to supplement their own food supply, and it is recorded that in 1534 excess stocks were replanted on Spanish soil. The first written record of the European potato can be traced to a receipt, dated 28 November 1567, for a shipment delivered to Antwerp from Las Palmas in the Grand Canaries, and at around the same time Spanish fishermen introduced the plant to the west coast of Ireland whenever they landed to dry their catch.

It is also recorded that Thomas Harriot (see 'The Telescope – and Why they Laughed at Galileo'), who was funded by Sir Walter Raleigh, returned from the Americas in 1586 with potato plants for England, despite many claims that Sir Francis Drake famously presented them to Queen Elizabeth I. Either way, the real spread of the potato was through the Spanish Empire, which planted them for its armies across Europe, and along the way peasants adopted the crop as an alternative to grain, as over-ground grain stores were often pillaged, leaving them with nothing to eat.

The European reliance on the potato was slow to begin with as many considered them to be poisonous, but by the middle of the eighteenth century the governments of both Germany and France were encouraging farmers to harvest them as a cheap and reliable food source for their growing populations, not to mention their own armies. French King Louis XVI (1754–93) actively promoted the new crop and, by the time they had lopped off his head, the annual French crop of potatoes was soaring. It had, by then, become the staple diet of most of northern Europe. And this is when this simple vegetable began to change the course of human migration throughout the world.

By the mid-nineteenth century the humble spud was worth more to ordinary European people than all the gold and silver the conquistadors had transported from the Americas, and it was recorded as being the food source that was underpinning the Industrial Revolution. In Ireland one-third of all farmland was dedicated to the crop that was fuelling the population boom. The poorest farmers could feed their families for a year on one small potato crop and the milk of a single cow. But when the crop failed in 1845 the resulting famine cost the lives of over a million people. The Irish Famine (1845–9) led to the migration of half the remaining population to England, where they found work as labourers on the growing railway network and, more importantly, to

America where the relatively new nation was in desperate need of a workforce.

Similar scale potato-crop failures throughout Europe, along with religious persecution, signalled the beginning of the mass migration to the United States during the second half of the nineteenth century, which played such a vital part in that country becoming an economic powerhouse. Suggestions that the humble potato was responsible for the development of the United States of America may not, after all, be exaggerated.

How Wrong Can You Be?
'The Americans may be good at making fancy cars and refrigerators but that doesn't mean they are any good at making aircraft. They are bluffing. They are excellent at bluffing.' So said Hermann Göring, commander of the German Luftwaffe, in 1942.

WORCESTERSHIRE SAUCE: WHO WERE MR LEA AND MR PERRINS?

During the 1830s, in the cellar below a chemist's shop in Worcester run by Mr Lea and Mr Perrins, sat a barrel of spiced vinegar – made according to an old Indian recipe for a customer, an English general who had returned from a tour of duty in the Raj. But it had been considered inedible and so the old soldier never bothered to collect it. Instead it sat for some years until one day, when clearing out the cellar, the chemists were about to throw away the barrel but fortunately decided to taste the contents first. They discovered, to their astonishment, that the mixture had nicely matured and so Worcestershire sauce was born.

You do have to wonder who on earth would dare to taste something they already knew to be unpalatable, after it had been left sitting in a cellar for several years. This reminds me of one of the great natural discoveries: cow's milk. How exactly did early man, or woman, learn that cows could be hand-milked? And what did they think they were doing when they found this out? And who was the first to then taste it?

The exact ingredients of the Worcestershire sauce recipe are jealously guarded, but we know they include soy sauce and anchovies, which is why the mixture started fermenting, and improving, over the years. Worcestershire sauce is thus a modern, and rather more palatable, version of the Roman fish sauce, garum. In 1838 the first, iconically labelled bottles of 'Lea & Perrins, Worcestershire Sauce' were released to the general public, and Mr Lea and Mr Perrins subsequently made their fortune from their distinctive-tasting condiment. The sauce remains hugely popular around the world to this day, especially in China and Japan, where it is praised for enhancing umami – or 'savouriness' – the fifth basic taste, considered fundamental to their cuisine.

WHO IS ON THE MENU? (CULINARY INVENTIONS)
Culinary Espionage à la Diamond Jim
Nicolás Marguery (1834–1910) was a culinary legend. His restaurant, Au Petit Marguery, was one of the most popular in Paris during the nineteenth century, and the great and the good of French society were regularly filling his dining room. In fact, they still do, enjoying surroundings that have hardly changed since the famous chef was

there, creating dish after dish in his busy kitchens, almost 150 years ago.

Nicolás Marguery's cooking, particularly his Sole Marguery – served in a sauce of white wine and fish stock blended with egg yolks and butter – was renowned throughout Europe and America. French cuisine was the world leader and its chefs jealously guarded the recipes of their signature dishes. Sole Marguery was one such dish and the story of how it emigrated from France to America is an intriguing one, full of subterfuge, audacity and a level of commitment that is hard to imagine.

As the immigrants who introduced the hamburger and the hot dog to New York settled throughout America, so the economy expanded, mainly through the growing railway network. Jim Brady (1856–1917), a salesman for the Manning, Maxwell and Moore Railroad Company, was part of this. And he was highly successful; the sums that he made selling railway tracks throughout America and across the world were so immense that he began to invest in diamonds and other precious stones, earning himself the nickname Diamond Jim.

A larger-than-life character, Brady also had a prodigious appetite. He was rumoured to consume a gallon of orange juice, steak, potatoes, bread, flapjacks, muffins, eggs and pork chops – just for breakfast. A mid-morning snack might be three-dozen oysters and clams, followed by lunch of another three-dozen oysters, three stuffed crabs, four lobsters, a joint of beef and a salad. During the afternoon, Jim would down six sodas and a seafood snack before taking a nap, and then came dinner: thirty-six oysters, six lobsters, two bowls of green turtle soup, steak, vegetables and a pastry platter. But it didn't end there: after a customary trip to the theatre, where he would eat two pounds of glacé fruits, Jim then rounded off his day with a supper of half-a-dozen game birds and a couple of large beers.

And this was Diamond Jim's intake every single day. His friend, the restaurateur Charles Rector, owner of Rector's Restaurant on Broadway, New York, would import several barrels of extra-large oysters from Baltimore every day, just for Jim. It's hardly surprising, therefore, that Charles once described the railway magnate as 'the best twenty-five customers I have ever had'.

One afternoon, during a marathon eating session with friends, Diamond Jim began telling the group about the exquisite Sole Marguery he had been served at Au Petit Marguery during a recent business trip to Paris. Unable to explain the recipe to Rector, he teased his friend that he might have to find another restaurant where he could eat Sole Marguery. Rector resolved on the spot to be the first in New York to serve the dish and set about obtaining the recipe, by fair means or foul. So he summoned his son, George, from Cornell University and sent him to Paris to obtain the secret ingredients. The young man arrived in Paris fully aware that he could not walk into the restaurant as a complete unknown and ask for it, so he applied for a job as dish-washer to see what he could learn from the kitchen staff. It soon became obvious that he would discover nothing from the chefs, who carefully guarded their recipes from all the menials in the kitchen, so George applied for a job as an apprentice chef.

It took him over two years of hard work to become senior enough to have the recipe for the legendary Marguery sauce explained to him, and as soon as it was he resigned and jumped on the first boat back to New York. It is said that both his father and Diamond Jim

Brady were waiting for him at the dockside as the ship drew in, with George shouting to them from the deck: 'I got it!' The young chef was sent straight into the kitchen to prepare Sole Marguery and, as legend has it, when Diamond Jim tasted the dish again he declared: 'If you poured this sauce over a Turkish towel, I believe I could eat all of it!' And that, dear reader, is how the famous dish of Sole Marguery à la Diamond Jim was invented.

Jim Brady died in his sleep in 1910 and only then did doctors discover he had an unusually large stomach, almost six times the size of that of a normal man. George Rector went on to take over his father's restaurant business. He also wrote cookbooks and cookery columns for newspapers, as well as hosting a radio show called *Dine with George Rector*. He is said to have literally dined out for the rest of his life on the story of how he had claimed the Marguery sauce for America.

Did an Old Card Player Invent the Sandwich?

Sandwich is, incredibly, not really even a proper word. It is a proper name, however. The village of Sandwich, first recorded in AD 642, is a picturesque and historic place in Kent; its name evolved from the Old English words *sand* and *wic*, meaning 'sand village' or 'town on the sand'. Now two miles from the coast, it was once a thriving seaport – the first captive elephant was landed there in 1255 before being delivered as a gift to Henry III. It was also the home of King Charles II's naval fleet under the command of Sir Edward Montague. When, in 1660, his grateful king made Montague an earl, the latter pondered which of the great ports he would honour with his new title. Bristol was one option and Portsmouth another, but the naval commander eventually settled for Sandwich and so his hereditary title became the Earl of Sandwich.

To date there have been eleven earls but the most famous of them, the one who invented every packed lunch in the Western world, was number four. John Montague (1718–92) was, like his great-grandfather, First Lord of the Admiralty, but unlike him he was both corrupt and incompetent. The Navy was in a state of complete disarray by the time it was called into action during the American Revolutionary War (1775–83); the eventual defeat of the British forces was regarded by many to be his fault. And that is hardly surprising as the Earl was far more interested in his life outside work, particularly gambling. Indeed, it was this that gave rise to the great culinary legend forever associated with him. According to the famous story, in 1762 he was playing cards with friends long into the wee small hours. Drunk and on a winning streak, Sandwich decided he needed some food and ordered waiters to bring him some meat but 'between two slices of bread'. This was to prevent his fingers from becoming greasy and marking the cards, helping his opponents to figure out his gaming pattern. The strategy worked and the snack soon caught on at the great gaming tables and gambling clubs of England, the 'sandwich' quickly becoming part of the English way of life.

It didn't help his reputation that Sandwich was also a member of the notorious Hellfire Club, a gentlemen's society set up to ridicule organized religion. No one knew what went on at their meetings – members didn't discuss them – but rumours were rife of orgies and satanic rituals. It was reputedly at one of these meetings that he became the victim of one of history's great one-liners. Sandwich is said to have abused Samuel Foote (1720–77): 'Sir, I don't know if you will die on the gallows or of the pox.' To which Foote shot back: 'That, my lord, depends on whether I embrace your principles or your mistress.' The story was speedily spread around London by Sandwich's many enemies.

How Wrong Can You Be?
In October 1969 Margaret Thatcher announced: 'It will be years, and not in my time, before a woman will become British Prime Minister.' In 1979 she proved her own political judgement to be out of touch with reality, and not for the last time either.

By the time of his death in 1792, Sandwich had become the most unpopular man in England. Even friends suggested his epitaph should read: 'Seldom has any man held so many offices and yet accomplished so little.' Yet the sandwich is not his only legacy to history. As First Lord of the Admiralty, Sandwich was one of the sponsors of the voyage Captain James Cook (1728–79) made to the New World in 1778. On 14 January, Cook became the first European to visit the Hawaiian Islands, which he originally called the Sandwich Islands in honour of his benefactor. Although the islands changed their name a century later, the South Sandwich Islands and the Sandwich Straits still bear the name of the old gambler, and inventor of the sandwich, to this day. Not to mention the expression it has given rise to, in the sense that we can now find ourselves sandwiched between two objects or two business appointments. I'm just relieved that the first earl selected Sandwich for his title. I'm not sure I'd fancy a cheese and chutney bristol or a corned beef and tomato portsmouth. Would you?

Creating Margherita

Margherita is possibly the most famous Neapolitan pizza in the world and is made with a topping of tomato, mozzarella, basil and olive oil. And it links an Italian queen with her country's poorest city. Margherita Maria Theresa Giovanna of Savoy, born in Turin on 20 November 1851, was the daughter of Ferdinand, Duke of Genoa, and his wife, Elizabeth of Saxony. With such a privileged background, it is hardly surprising that Margherita's future was planned long in advance, and on 21 April 1868, aged just sixteen, she was married to Umberto, the heir to the Italian throne. Margherita became Queen in 1878 when Umberto succeeded as the second King of Italy, then a newly unified country. Prior to that Italy was instead fragmented and populated by principalities and kingdoms. The new queen's passionate patronage of the arts and outspoken support for organizations such as the Red Cross earned her the respect of the young nation.

Indeed, she was held in such affection that the third-highest mountain in Africa was named after her, Margherita Peak (which you could translate as 'Mount Daisy' as that is the meaning of the name in Italian), and she is also commemorated in a culinary way. It was a visit by the popular queen to Naples in 1889 that prompted Raffaele Esposito, owner of the Pizzeria di Pietro, to prepare a special meal in her honour. Using the colours of the new national flag – green, white and red – Esposito combined cheese and tomato (white and red) with basil

87

(green) to create what has become one of the world's biggest-selling pizzas, and the combination provides the base ingredients for most of the others. He called it pizza Margherita (daisy pizza, in other words, although that doesn't sound quite as appetizing) after his queen. A culinary invention we have all enjoyed at some point or another.

Classic Invention from the Thousand Islands

The exotically named Thousand Islands are a cluster of islands (there are actually 1,793 of them) in the Saint Lawrence River on the border between America and Canada. Every July and August, when New York turns into a humid oven, city dwellers traditionally escape to the islands, where many of the houses are owned by these vacationers. At around the turn of the twentieth century, a well-known Thousand Island fisherman, George LaLonde Jr, was teaching a prominent New York actress, May Irwin (1862–1938), to fish. One evening, following a fishing expedition, LaLonde's wife Sophia served one of her 'shore dinners': Irwin was particularly impressed with the salad dressing, made from mayonnaise and tomato ketchup mixed with finally chopped green olives, pickles, onions and hard-boiled eggs. Impressive as it was, the dressing was essentially made with whatever came to hand.

On these islands at that time, with little access to fresh ingredients, dishes had to be prepared using store-cupboard basics. May Irwin asked for the recipe and immediately passed it on to her friend and fellow Thousand Island vacationer George C. Boldt (1851–1916). Boldt, owner of the Waldorf Hotel in New York, was equally impressed with the sauce, asking his maître d',

Oscar Tschirky, to refine it and introduce it to the diners at his hotel. Today, Thousand Island dressing is internationally known and has itself become a store-cupboard basic, sold in jars in supermarkets throughout the world. A pure invention from a Thousand Islands fisherman's wife.

The Inventors of Breakfast Cereal

The brightly coloured cereal packets on your breakfast table, full of chocolate and sugar and covered in cartoon elves and grinning tigers, are in fact the unlikely last remnants of a bizarre, long-running battle that raged in nineteenth-century America between an equally unlikely set of combatants: vegetarians, water-cure fanatics and the Seventh-day Adventist Church.

It was all powered by a growing obsession with regulating bodily functions. At the time, most Americans ate an English-style cooked breakfast, which was a substantial meal, heavy on pork and other meats and very low in fibre. As a consequence, many suffered constipation and other painful gastric disorders.

But, being the nineteenth century, nothing happened by halves. The first spokesman in the health-food revolution was the Reverend Sylvester Graham (1794–1851). A vegetarian with no medical training, he was determined that wholemeal flour was the answer, and his very profitable Graham bread and Graham crackers were the result of his efforts. Vegetarianism and temperance became wildly popular for a while; meat-eating was declared to be unhealthy, not to mention a catalyst to equally negative carnal desires, and coffee and tea were both condemned as poisons. It wasn't long before Graham's supporters were declaring that their search for 'healthy' substitutes based on grains and cereals was for the common good, some no doubt recognizing how this might earn them very healthy sums of money in the process.

In 1858, Dr James Caleb Jackson (1811–95) took over an unsuccessful water-cure resort in New York, renaming it 'Our Home Hygienic Institute'. Patients were subjected to a punishing regime of baths and unpleasant treatments and fed a restricted diet based on various grains, similar to the way farm animals were. In 1863, Jackson created the first breakfast cereal, which he called Granula, but it was hardly fast food; it had to be soaked overnight in milk before it was even possible to chew through the stone-hard crumbs. Even so, Granula became very popular, earning Jackson ten times the cash he had invested in developing it.

Meanwhile, in Battle Creek, Michigan, the Seventh-day Adventists were running a health institute, the Battle Creek Sanitarium, where the latest in dietary reform was being introduced. But it didn't really catch on until John Harvey Kellogg (1852–1943) was put in charge. Dr Kellogg had been hand-picked for the job as his medical and spiritual training had been supervised at every stage by the Adventists. Following his experience of living in a boarding house during training, where cooking was impossible and restricted to the vegetarian diet required by his religion, the hungry young man recognized the need for a ready-cooked breakfast cereal that required no

---

With over fifty foreign cars already on sale here, the Japanese auto industry isn't likely to carve out a big slice of the US market.

Business Week USA, 2 August 1968

---

preparation. In 1880, he came up with a mixture of wheat, oat and maize meal baked in little biscuits, which he cheekily called Granola, and it became a huge success, in the land of overnight successes.

A few years later, in 1893, a Denver lawyer called Henry D. Perky (1843–1906) invented a completely different product to cure his indigestion, calling it Shredded Wheat. The wheat was steamed until thoroughly softened, then squeezed between grooved rollers to form strands that were then pressed together and cut into biscuits, referred to by Perky as 'my little wholewheat mattresses'. Unfortunately, the process didn't work as the moist wheat biscuits soon perished. Kellogg then went to see the disillusioned inventor and offered Perky $100,000 for the patents he had taken out for manufacturing his cereal, but he lost his nerve and retracted the offer. He would later regret this, however, particularly since during their conversation he had shared his secret of how Kellogg products were dried by slow heating so that they remained in perfect condition for a long period of time. Armed with this knowledge, Perky tinkered with his machinery, began to dry his Shredded Wheat and then sat back to watch the dollars roll in, becoming an immensely rich man in the process.

Kellogg was naturally envious and, after a long period of experimentation, came up with a process in which wheat was cooked, flattened into flakes and then dried. Granose Flakes, as he called them, would soon prove to be a significant commercial discovery, but not for the good doctor. With no real head for business, he was mostly interested in his sanatorium, and for a while his patients were the only people who could buy his products.

The man primarily responsible for speeding breakfast cereal out into the grocery stores of the nation was Charles William Post (1854–1914). He entered the cereal business after a string of entrepreneurial failures that had led to a physical collapse. As a patient at Kellogg's sanatorium in 1891, he didn't find a cure, but he did come to realize that health foods and, in particular, coffee substitutes were potential goldmines. The idea alone must have been enough to cheer him up a bit. After leaving the sanatorium, he started his own health institute in Battle Creek and within four years he had developed Postum, a wheat-and-molasses based hot drink. Using everything he knew about sales, Post then mounted an advertising campaign and his product became a success.

There was, he said, no limit to the number of physical and moral ills (even divorce or juvenile delinquency) that were caused by coffee, but it could all be banished with Postum, the beverage that promised to 'make the blood redder'. Two years later, he launched what would prove to be an even bigger hit. Grape Nuts was a failure as a grain beverage, as it was originally marketed, but once it was rebranded as a breakfast cereal it quickly became a bestseller. (It was sweetened with maltose, which Post called grape sugar and which he thought had a nut-like flavour; hence the name.) By 1902, Post was making over a million dollars a year, a lot of money even now but a fortune in those days.

J. H. Kellogg's younger brother and general office assistant at the sanatorium, Willie Keith (1860–1951), followed with an improvement on the Granose idea – flakes made from corn. Eventually, the two Kellogg brothers fell out and in 1906, W. K. (whose signature still appears on every cereal packet as the company's trademark today) founded the great Kellogg breakfast food empire with his toasted Corn Flakes. Originally called the Battle Creek Toasted Corn Flake Company, it was renamed the Kellogg Company in 1922, while the product on which it was founded became what must be the most celebrated breakfast cereal in the world. At the time of its conception, hundreds of other would-be cereal pioneers had leapt into the field as well, many journeying to Battle Creek itself to start their businesses. Soon thirty different cereal-flake companies, most of them fly-by-night operations, had crowded into the small town. And Americans had scores of cereals to choose from, each promising to cure their every ailment.

But despite their origins in the health-food movement, breakfast cereals have no special nutritional value beyond the food value of the grain from which they are made. Which is why many of them are now artificially supplemented with extra vitamins. In fact, it is the milk they are eaten with that provides most of the nutrients they otherwise lack.

A Crisp Answer to a Complaint

On 24 August 1853, George Crum (born George Speck, 1822–1914) was working as head chef at Moon's Lake House in Saratoga Springs, New York, when a customer complained that his French fries were too thick and not 'as they should be'. Crum was so annoyed by this remark that he decided to exaggerate his response by slicing the potatoes as thinly as he possibly could and then frying them in hot oil. To his great surprise, the customer was delighted and Saratoga potato chips, or crisps as we know them in England, proved to be so popular with other diners that Crum was soon able to open his own restaurant with the profits he made from his invention.

Which Caesar Invented My Salad?

The name, of course, conjures up the image of a grumpy, old toga-clad emperor tucking into a spot of lunch before perhaps throwing a Christian or two to the lions, in the name of entertainment, in front of the people of Ancient Rome. But instead it turns out that the Caesar salad was invented less than a hundred years ago and emerged from the most unlikely of places: Mexico.

Caesar Cardini (1896–1956) was born in Italy and emigrated to America with his three brothers at the beginning of the First World War. It was the era of Prohibition, also known as the Noble Experiment, when between 1920 and 1933 the sale, manufacture and transportation of alcohol were banned in the United States. The aim was to improve the morality and behaviour of the American people, but in fact it just encouraged a huge rise in organized crime. Realizing the lengths people would go to just to have a drink or two, Caesar and his brother Alex saw a legal business opportunity and seized hold of it. In 1924, they moved the short distance across the Mexican border from Los Angeles and set up a restaurant in Tijuana, after the town became a firm favourite for southern Californians looking for a weekend party.

The Cardinis' combination of strong alcohol and tasty Italian food proved to be a winner and their Fourth of July celebrations were so oversubscribed that, according to Caesar's daughter Rosa, her father soon ran out of ingredients to feed his drunken customers. He responded by throwing together a salad of basically whatever he had

94

left in the kitchen: lettuce, croutons, Parmesan cheese, eggs, olive oil, lemon juice, black pepper and Worcestershire sauce. Perhaps trying to make up for his dish's simplicity, he brought the salad to the table and, with a theatrical flourish, tossed it in front of his customers so that every leaf was covered in the thick dressing.

The story goes that the dish proved to be so popular with a group of partying Hollywood film stars who had flown in for the weekend that Alex named it 'Aviator Salad' in their honour. Later, once their restaurant was established on the ground floor of the Hotel Commercial, the Cardinis could afford to admit the truth about its origins and the salad was renamed after Caesar. It continued to be a firm favourite among the stars of the day, who demanded the dish wherever they travelled in the world. Thanks to his impromptu way of dressing a salad, Caesar Cardini became a rich man and eventually trademarked his famous creation, in 1948. Today the Cardini Company remains America's favourite producer of an ever-growing range of oils and dressings.

How Wrong Can You Be?
Newsweek magazine, when predicting popular holiday destinations during the mid-1960s, advised: 'For the tourist who really wants to get away from it all then there will be safaris in Vietnam.'

Why Chicken Kiev Rules Supreme

Possibly the most famous chicken recipe in the world, what we know as chicken Kiev was originally an Italian dish called *pollo sorpresa*, or 'chicken surprise'. The surprise is the molten garlic butter that jets out of the breadcrumb-covered chicken breast once you stick a fork in it. The French version of this – acknowledging the fact that the whole point of a surprise is that you don't know it's coming – is known as *suprême de volaille*, 'best chicken' (although it can also refer to chicken breast cooked in a rich white sauce). It was the French version that gave the world the taste for the dish, and it was all due to Napoleon Bonaparte (1769–1821). He famously remarked that 'an army marches on its stomach', and he once offered a prize of 12,000 francs to anyone who could devise a method of preserving food to help him keep his troops on the move. After fourteen or so years of experimentation, it was Nicolás Appert (1749–1841) who won the prize, in January 1810, for his technique of pre-serving food in vacuum-sealed bottles.

It is said that one of the first foods Appert managed to preserve in this way was his version of chicken supreme, and as a result the dish was exported around Europe at a speed never previously known. His book *L'Art de Conserver les Substances Animales et Végétales* ('The Art of Preserving Meat and Vegetables') was published in the same year and was the first cookbook on modern preser-vation methods. Within ten years, canning had evolved, based on the technique established by Appert, who was known thereafter as the 'father of canning'. His method was described as 'appertization'.

But how did chicken supreme later become chicken Kiev? According to the Russian food historian William Pokhlebkin – his surname deriving from *pokhlebka*, or 'stew', the underground nickname adopted by his father during the 1917 revolution – the Russian version of the recipe was invented in the Moscow Merchants' Club in the early twentieth century. This was at a time when communist Russia rejected everything outside its own borders and only Russian names were tolerated. The canny chef at the club renamed the chicken supreme dish chicken Kiev, and it became hugely popular as a result. The twentieth century also saw a huge wave of Russian emigrants escaping persecution in their homeland. Many travelled to America, and restaurants, particularly on the East Coast, began to call chicken supreme chicken Kiev to attract new customers familiar with the dish from their native land. It was during the two world wars that the new name migrated back towards Europe. In 1976, chicken Kiev made history by becoming the first ready meal produced and sold by Marks & Spencer, another step in a different revolution – the conquest of the kitchen by fast food.

An ironic footnote to the overall story: despite Nicolás Appert's success in inventing a process directly leading to the mass production of tin cans, it would be nearly fifty years before another inventor came up with the idea of a can opener. Now, that's worth thinking about for a moment (see 'The Tin Can').

POPULAR CULTURE

THE HULA HOOP

The earliest record of humans using hoops, in this case fashioned from vine, was made during the fifth century BC and attributed to the Greeks, who were known to use them for both exercise and leisure. Since then, children and adults alike have been rolling, spinning, throwing, catching and anything else they could find to do with them.

As previously mentioned, when the English explorer Captain James Cook sailed into Hawaii in January 1778 he initially declared them to be the Sandwich Islands, in honour of the First Lord of the British Admiralty at the time, the 4th Earl of Sandwich who, incidentally, was the sleazy and incompetent old gambler who invented my lunch today. By the way, it was also Sandwich who is generally thought to be responsible for Britain losing the American War of Independence as he insisted on keeping the most powerful navy in the world in European waters to keep an eye on the French, instead of sending ships to the colonies. But I digress. After Cook had set his anchor and rowed ashore in search of fresh supplies, one of the many observations he made in his extensive records of what was to be his third and last tour of the Pacific was of the ancient and traditional dance of '*hula kahiko*' that had been passed down through the generations of the

Polynesian people who inhabited Hawaii. It became his last voyage because the same, once friendly, people killed him when he returned to Hawaii the following year as he made his way back to Europe.

Nothing much more was notable about the traditional dance of the Pacific Islands people until 1865 when Paul Iria and Ken Vezina began manufacturing plastic hoops, although they were not a great success commercially. That is until American toy manufacturer Arthur 'Spud' Melin returned from a vacation to Australia in 1957 and told his business partner Richard Knerr of the people he had encountered exercising with plastic hoops. They soon made a few cheap prototypes and gave them away to children in the area to see what they would do with them. By the June of the following year, and after a national marketing campaign funded largely by the successes they had enjoyed with their Frisbee product (launched the previous year after the pair had watched people spinning plastic picnic plates to each other on a beach), the hula hoop was ready for distribution.

How Wrong Can You Be?
Stan Smith (b. 1946) was rejected as a ball boy for a Davis Cup tennis match as he was considered to be too clumsy and awkward around the court. He went on to play in three Grand Slam finals, winning two of them, and was champion of eighty-seven other tournaments around the world in a thirteen-year career. He also won eight Davis Cups.

Never in the history of manufacturing (until perhaps

the iPod was released over half a century later) has there been a more instant success, with over twenty-five million hula hoops being shifted throughout America alone, at $1.98 each, during the first four months of sales. A new American craze had been born and we can only guess at what may have happened for British industry if Captain James Cook had managed to escape the Hawaiian Islands with his life intact nearly two centuries earlier and brought home news of the hula-hoop dance.

THE RECORD EXECUTIVE WHO MIGHT HAVE KICKED HIMSELF TO DEATH

If it was all about fairy tales then there can be none bigger than the story of the four unpolished teenagers from post-war Liverpool who went from having a handful of chords and a few third-hand guitars to becoming the biggest and most influential pop band the world will ever see.

In July of 1957, at the ages of fifteen and sixteen respectively, Paul McCartney and John Lennon met for the first time at a summer fete in a suburb of Liverpool. The budding guitar enthusiasts were soon playing together and formed their own band, which, in January 1958, McCartney invited his school friend George Harrison to join. Together with guitarist Stu Sutcliffe and drummer Pete Best, the band were soon performing in pubs and clubs around the city, and by 1961 had even completed a couple of three-month residencies at a club in Hamburg.

Despite their confidence and self-belief, the young band had no idea how to even approach anybody within the London music industry, let alone secure a recording contract. That was until a Liverpool record-shop owner,

Brian Epstein, who was himself only in his mid-twenties, started noticing the band's name on posters around the city and read a feature article about them in a local music fanzine. Finally, after the Beatles had recorded their first demo, a song called 'My Bonnie', and a customer walked into Epstein's store asking for a copy, his curiosity led him to the now famous Cavern Club in Liverpool where the Beatles were playing a lunchtime show on 9 November 1961.

Conscious of their growing reputation, Brian Epstein was eager to be involved with the band and felt he could use his London record-distributor contacts in their favour. He was also suitably flattered that the Beatles, all being regulars at his record shop, actually recognized him. They too were aware that his contacts could be to their advantage and an agreement was quickly reached for Brian Epstein to become the manager of the Beatles. Although a formal contract wasn't signed until 24 January 1962, Epstein went to work immediately and arranged for demos to be recorded that were hurried down to London and the offices of EMI, Decca Records, Columbia, Pye and Philips.

Luck was on their side, as at Decca Records a young assistant called Mike Smith had seen the band perform at the Cavern only few weeks earlier and had been impressed by the reaction of the audience more than he had by their music. As a result, Epstein was invited to take the band to an audition at Decca Studios in London for Smith's boss, senior A & R man Dick Rowe, on 1 January 1962. On New Year's Eve the Beatles squeezed into the back of a small van for what turned out to be a ten-hour journey through the snow, to arrive in London at 10 p.m., just as the New Year celebrations were gathering pace in the capital city.

Unsurprisingly, the four youngsters were only too happy to join the party and by the time of the audition, at 11 a.m. the following morning, none of them were feeling particularly lively. Mike Smith himself arrived both late and hungover and immediately informed the group that their own equipment was sub-standard and they would have to use amplifiers provided by the studio. That meant they could have travelled by train instead. For the one-hour audition Epstein had chosen fifteen songs from their live set list, which included only three original Lennon and McCartney tunes. All were nervous in the unfamiliar surroundings and drummer Pete Best apparently played the same beat on every song. John Lennon, who was supposed to be the main singer, left most of the vocal duties to Paul McCartney and even guitarist George sang on three of them whilst Lennon confined himself to the background.

Even so, everybody was happy enough with the audition and left believing the deal was as good as done. After celebrating in a north London restaurant the group squeezed back into the van for the journey north and the long wait for a decision. Three weeks later Brian Epstein telephoned Decca Records and spoke to Dick Rowe, confident of a positive response. Instead Rowe bluntly told him, 'Groups with guitars are on the way out.' Epstein was shocked as Rowe continued, 'The Beatles have no future in show business. You have a decent record business going up there, why don't you go back to that?' Epstein recovered his composure and replied, 'You must be out of your mind; one day these boys are going to be bigger than Elvis.'

Within two months Brian Epstein and the Beatles had their recording contract with EMI subsidiary Parlophone and at the close of 1963 were the biggest selling act in British recording history. By the end of their short career the Beatles had sold hundreds of millions of records around the world and were indeed 'bigger than Elvis'. Dick Rowe, on the other hand, has gone down in show-business history as 'the man who turned down the Beatles'. However, for the rest of his life Rowe denied this version of events. He claimed that during the same New Year's Day auditions another unsigned band, Brian Poole and the Tremeloes, had also performed and Rowe had said to Smith, 'I like them both, you will have to choose.' He then went on to claim that Smith selected Brian Poole and the Tremeloes as they were 'from London and would be easier to work with'.

Despite the Beatles's initial disappointment, there did not appear to be any animosity between them and Rowe as soon after they had become superstars the Decca man bumped into George Harrison in a television studio and, instead of sarcasm, received a tip off about an upcoming young band called the Rolling Stones, whom he duly signed. John Lennon, on the other hand, when asked if he thought 'the man from Decca would be kicking himself', replied, 'Yeah, hopefully to death.' Even so, Lennon and McCartney wrote the Rolling Stones' first hit with Decca Records, 'I Wanna Be Your Man', which charted at number 12 in November 1963.

However, there is also strong evidence that Dick Rowe's judgement was usually reliable as he also had a hand in signing the Moody Blues, Tom Jones, the Small Faces, the Animals, the Zombies and Van Morrison to Decca Records. It is worth noting that the Tremeloes achieved moderate success with the label, peaking at number 2 in the UK charts with a cover of Roy Orbison's 'Candy Man' in 1964. Some years later the Beatles producer George Martin defended Rowe and stated that he too would not have signed the group on the evidence of the Decca audition tapes he had heard.

> It is an idle dream to imagine that automobiles will take the place of railways in the long distance movement of passengers.
>
> American Railroad Congress report, 1913

THAT'S A TERRIBLE IDEA FOR A BOOK

The truth is that almost every successful book you have ever read has been repeatedly rejected by publishers (unless, of course, it was written by an established author with a ready-made audience) who may look back on a missed opportunity. So that means there is enough material to produce an entire volume on this subject alone, although in this case we will keep the examples down to some of the most famous with interesting tales to tell.

Animal Farm by George Orwell

The book that quickly became recognized as a modern classic was rejected four times before finally being accepted and published by Secker and Warburg in 1945. Despite George Orwell (1903–50) being an established and popular writer at the time, it would take him eighteen months before he could find a publisher who would consider releasing *Animal Farm*. The problem centred on its content, which was cynically anti-communist and therefore, at that time, anti-Russian. Orwell even had his main character, which was clearly modelled on Stalin himself, depicted as a fat, wheezing pig.

In the middle of the Second World War, when Russia was an important ally of Britain and the United States in the fight against the Nazi Party, nobody in London or New York was particularly keen to be associated with *Animal Farm*. Anti-Soviet literature was something to be avoided at all costs and even his regular publisher, the left-leaning Victor Gollancz, turned the book down. Faber & Faber and Nicholson & Watson were the next to pass, and Jonathan Cape, which did accept the first draft, was then warned off by Peter Smollett, the head of the Russian section of the British government's Ministry of Information. Smollett was later revealed to be a Soviet spy.

Cape withdrew its offer and explained to Orwell, 'If the fable were addressed generally to dictators and dictatorships at large then publication would be all right, but the fable does follow, as I see now, so completely the progress of the Russian Soviets and their two dictators (Lenin and Stalin), that it can only apply to Russia, to the exclusion of the other dictatorships. Another thing: it would be less offensive if the predominant caste in the fable were not pigs. I think the choice of pigs as the ruling caste will no doubt give offence to many people, and particularly to anyone who is a bit touchy, as undoubtedly the Russians are.'

In his London Letter column for *Partisan Review* on 17 April 1944 Orwell complained that it was 'now next door to impossible to get anything overtly anti-Russian printed. Anti-Russian books do appear, but mostly from Catholic publishing firms and always from a religious or frankly reactionary angle.'

The war was over by the time *Animal Farm* was finally released, but even then the publisher, Fredric Warburg, faced pressures from his staff and even from his wife who did not believe it to be a good idea to appear ungrateful to Russia and the 'glorious Red Amy'. Warburg forged ahead anyway and released the book that went on to sell over twenty million copies and to be included at number 31 on the Modern Library list of best novels of the twentieth century.

In a classic case of life imitating art, the internet giant Amazon withdrew the Kindle editions of *Animal Farm* and Orwell's classic follow-up to it, *Nineteen Eighty-Four*, when, on 17 July 2009, doubts emerged over the publisher rights to the titles. In a classic Big Brother move, Amazon remotely deleted all copies on customer devices and provided refunds. There was an international outcry and Amazon spokesman Drew Herdener was forced to announce that the company was 'changing our systems so that in the future we will not remove books from our customer's devices in these circumstances'.

Zen and the Art of Motorcycle Maintenance by Robert Pirsig

I once read about the managing director of an international publishing firm who admitted, whilst addressing a publishing conference in the United States, that he had no idea how to single out a potentially bestselling book simply from the submissions his company received. And he went on to add that he didn't believe anybody else had a clue either. To some extent this is true. There is no way of knowing, in advance, what members of the book-buying public are all going to like, at the same time. Or which books the important book reviewers will choose to take home with them from the hundreds they receive every day. Or which books the popular television shows will want to talk to the author about, or which writers will be asked to talk about their efforts on the radio shows. And, of course, there is no way of knowing if all of these important factors may happen to line up at the same time six months after the writer has handed their manuscript in to the publisher, if he or she ever gets that far.

Nearly all successful writers will have been rejected by somebody somewhere before they are eventually considered to be worth publishing, and even then they will only be considered in the context of their previous book, if there was one. And this is why Robert M. Pirsig's debut novel *Zen and the Art of Motorcycle Maintenance* is worth mentioning here. Pirsig (b. 1928), a university professor with a nervous breakdown and a spell in a psychiatric hospital on his CV, was forty-six years old when he published *Zen and the Art of Motorcycle Maintenance* in 1974. It can only be a credit to the author, and an example to all aspiring authors, that *Zen* was famously rejected 121 times by American publishers, a feat that is listed in *The Guinness Book of Records* as the bestselling book that has been subject to the most rejections. That is probably not a world record of which Pirsig is particularly proud, but he will certainly be proud of the subsequent five million sales that he and the publisher who eventually accepted it, William Morrow & Co., achieved.

How Wrong Can You Be?
Variety magazine considered the rock 'n' roll phenomenon of the mid-1950s and then declared: 'It will be gone by June.'

Carrie by Stephen King

When the unknown writer Stephen King (b. 1947) started work on his fourth book he was a high-school English teacher living in a trailer with a disconnected telephone and only his wife's second-hand typewriter to work on. Having suffered the repeated rejection of his first three books King was losing heart and, at one point, gathered up the pages of his manuscript and threw them away. When his wife Tabitha realized what he had done she retrieved the pages, reviewed them, put them back in front of him and encouraged him to continue. King himself later recalled, 'I persisted because I was dry and had no better ideas. My considered opinion was that I had written the world's all time loser.'

With the manuscript completed the writer again suffered thirty rejections before receiving a telegram from Bill Thompson. An editor at Doubleday, Thompson had tried telephoning King only to find he had removed the telephone line in order to cut down on expenses. The message read, 'Carrie officially a Doubleday book – $2,500 advance against royalties. Congrats Kid, the future lies ahead, Bill.' The future certainly did lie ahead as *Carrie* sold a reasonable 13,000 copies before King and Doubleday received an offer of $400,000 for the rights to a paperback version, which they split between them. By the end of the year *Carrie* had sold over a million copies, and King would become one of the bestselling novelists of the century. Carrie went on to sell over five million copies and has spawned three films and a theatre production. More importantly for the author, it launched a career that has so far produced fifty books, numerous films and worldwide sales in excess of 350 million.

On the Road by Jack Kerouac

When *On the Road* was first published in 1957 it was described by the *New York Times* as 'the most beautifully executed, the clearest and most important utterance so far made by the generation Kerouac himself named years ago as "beat" and whose principal avatar he is'. After dropping out of college Jack Kerouac (1922–69) moved to the Upper West Side in New York City where he met William Burroughs, Herbert Huncke, Alan Ginsberg, Neal Cassady and John Clellon Holmes. All of them were writers, some published and others still waiting for a break. In 1942 Kerouac joined the United States Merchant Marine and, whilst at sea, wrote his first book, *The Sea Is My Brother*. The young writer described it at the time as a 'crock of shit' and never even tried finding a publisher for it. It would be seventy years, forty-two of them after his own death, before Kerouac's debut effort could be found on the bookshelves.

On returning to New York, Kerouac worked on a second book with William Burroughs called *And the Hippos Were Boiled in Their Tanks*, which was also never published during either man's lifetime. It was Burroughs's first attempt at novel writing and he would later become the first of the group to gain notoriety when his book *Junky* became a controversial bestseller in 1953. Ginsberg's debut collection of poems, *Howl*, made him famous in 1957 when it became the subject of an obscenity trial; Herbert Huncke achieved moderate success in 1964 and became more productive during the 1980s; John Clellon Holmes produced a novel called *Go* in the 1950s and again was more productive later in his career; and Neal Cassady, who was regarded as the brightest member of the Beats and was the model for the central character in

On the Road, wrote next to nothing at all apart from letters and an autobiographical novel in 1971.

Kerouac, on the other hand, who had taken a road trip across America with Cassady in 1946, had been prolific and had completed four full-length novels before sitting down to write the first complete draft of *On the Road* in 1949 in, by his own estimation, only three weeks. He then spent another three weeks preparing the final draft. Meanwhile, his first novel to be published, *The Town and the City*, was released in 1950 and, although it received good reviews, the book sold poorly. Kerouac was more confident about *On the Road* but found the response from the publishers he approached to be less than enthusiastic. The book was repeatedly rejected on the grounds that most editors were uncomfortable with the idea of publishing a book that was sympathetic towards marginalized groups (the Beats) in post-war America and were concerned that Kerouac's graphic descriptions of drug use and homosexuality could result in obscenity charges. Kerouac reached a dead end and doors closed all around him. When his wife, Joan Haverty, left him after she had fallen pregnant, the young writer went back on the road and found part-time manual work to fund his travels. For the next five years Kerouac continued to submit *On the Road*, without success, and wrote what would later become the draft manuscripts for ten more novels. He also fell into bouts of depression linked to his heavy drug and alcohol abuse.

Finally, in 1957, Viking Press made an offer for *On the Road* but demanded major revisions to which Kerouac reluctantly agreed. After all of the sexually explicit passages were removed, and all the character names had been changed for fear of libel, *On the Road* was ready for publication. In July 1957 Kerouac moved to Orlando in Florida to await the release, a full seven years after the book had been completed. Within weeks, Kerouac woke up one morning to read a review of *On the Road* written by Gilbert Millstein of the *New York Times* that proclaimed, 'Just as, more than any other novel of the 1920s *The Sun Also Rises* came to be regarded as the testament of the Lost Generation, so it seems certain that *On the Road* will come to be known as that of the Beat Generation.' *On the Road* gave Kerouac instant fame, although those describing him at the time as an overnight success were woefully inaccurate.

Publishers were now beating a path to his door and his other previously unwanted and rejected manuscripts were eagerly snapped up. *On the Road* became such a famous book during 1957 that within nine months of its release Kerouac no longer felt safe in public after being badly beaten up by three men in New York. The book changed the lives of the friends he fictionalized in the story and Neal Cassady was regularly arrested and searched for drugs. There were as many people in conservative post-war America who were disgusted by the exploits of Kerouac and his characters as there were young Americans who could relate to them.

Kerouac's star shone brightly after the release of *On the Road* but his lifestyle cost him his health. In 1969, at the age of forty-seven, Kerouac was sitting in his favourite chair drinking whisky and recovering from yet another bar fight two weeks earlier, when he suffered an internal haemorrhage as a result of a lifetime as a heavy boozer. He died the following morning having never regained consciousness.

The multiply rejected *On the Road* went on to become a genuine modern classic, and since its publication in 1957 it has never sold fewer than 60,000 copies in any given year. In 1998 the Modern Library listed *On the Road* as number 55 on its list of the 100 best English-language novels of the twentieth century, and it also made *Time*'s similar list.

The Tale of Peter Rabbit by Beatrix Potter

The Tale of Peter Rabbit was written in 1893 by part-time artist and illustrator Beatrix Potter (1866–1943) for five-year-old Noël Moore, the son of Potter's former governess, Annie Carter Moore. Eight years later, in 1901, the author revised the stories and approached a number of publishers, in vain, before deciding to have a small number printed herself so she could use them as gifts for family and friends.

The following year, one of the publishers who had turned the book down happened, by chance, upon a copy of the privately printed edition and changed his mind. Frederick Warne & Co. contacted Potter and agreed to release a trade edition of *The Tale of Peter Rabbit* with colour illustrations, which by the end of 1902 had sold a very healthy 20,000 copies. Over the years that followed, *The Tale of Peter Rabbit* sold in excess of forty-five million copies and generated an industry that includes film, cartoons, toys, clothing, food and over twenty-five new characters that remain in print over a century later.

Chicken Soup for the Soul by Jack Canfield and Mark Victor Hansen

In 1990 a pair of motivational speakers called Jack Canfield (b. 1944) and Mark Victor Hansen (b. 1948) came up with the idea of compiling a book of inspirational stories to sell at their seminars. For centuries chicken soup has been regarded as a comfort food and has always been given to youngsters who are feeling unwell, particularly throughout Jewish communities. In fact, twelfth-century physicians and apothecaries would prescribe what was then known as 'Jewish Penicillin' to those with colds and flu. And with this in mind Caulfield and Hansen decided to call their book, which was intended to inspire the reader and to lift their spirits, *Chicken Soup for the Soul*. This was their first mistake, according to the publishers they approached with idea.

Canfield later recalled, 'The first time we went to New York we visited a dozen publishers in two days with our agent and nobody wanted the book. They said it was a stupid title, nobody bought collections of short stories, that there was no edge, no sex, no violence so why would anybody want to read it?' However, the pair remained undeterred and spent the next two years compiling 101 stories, with a little help from friends and associates, and continued approaching book publishers. At one point Hansen even began walking into meetings with a brief-case full of signed requests from as many as 20,000 customers who all guaranteed to buy a copy as soon as it was released. In all, it is estimated that they approached over a hundred publishers, all of whom turned them away.

That was until Heath Communications, a small and struggling independent publishing house who specialized in self-help books on subjects such as drug addiction and alcoholism, was given a copy and loved the idea. However, Heath was on the verge of going out of business and had little money to pay for the manuscript. So Caulfield and

Hansen signed the rights over for no fee at all, and instead agreed to share in the royalties, if there were any. As soon as the book was distributed *Chicken Soup for the Soul* became a *New York Times* bestseller and the first in a range of over 200 *Chicken Soup* titles, such as *Chicken Soup for the Teenage Soul*, *Chicken Soup for the Adopted Soul* and *Chicken Soup for the Celebrity Soul*, that have sold over 125 million copies and are translated into sixty different languages around the world. So that was a stupid title, was it?

How Wrong Can You Be?
Dr Paul Ehrlich has been prolific in his failed predictions. In a speech given during Earth Day in 1970 he confidently announced, 'In ten years all important animal life in the sea will be extinct. Large areas of coastline will have to be evacuated because of the stench of dead fish.'

Lolita by Vladimir Nabokov

By the time Russian author Vladimir Nabokov (1899–1977) set about writing what would become his seminal work, he had already fled Russia for Germany after the Bolshevik Revolution. Then he fled Germany for France at the outbreak of the Second World War, followed by France for America as the Nazis approached Paris few years later. He had written a number of novels in his native language and two in English, and was known as an established writer.

In the summer of 1953 Nabokov set off with his wife Vera on one of their regular butterfly-collecting holidays in the western part of the United States, where he used his free time to write the first draft of *Lolita*, with Vera acting as typist, editor, researcher, chauffeur, proofreader, agent, secretary and cook. When the Nabokovs returned home at the end of their trip, the author attempted to burn the unfinished drafts. It was Vera who prevented him.

Nabokov was persuaded to complete the manuscript, which he managed by 6 December 1953, and considered offering the work under a nom de plume, but Vera found the book was rejected by just about every publisher she approached. One of the rejection letters pointed out that the book was 'overwhelmingly nauseating, even to an enlightened Freudian. To the public it will be revolting. It will not sell and could do immeasurable harm to your growing reputation. I suggest you bury it under a stone for a thousand years.' Other publishers warned of obscenity trials and quickly rejected it. Ultimately six publishing houses rejected *Lolita* before Nabokov turned to his translator, Doussia Ergaz, with instructions to prepare the manuscript for the French market where he reasoned he would find less resistance. Finally, the book reached Maurice Girodias of Olympia Press, much of whose publishing catalogue was considered to be pornographic rubbish and, unsurprisingly, they agreed to produce the book.

Nabokov had been unaware of the publisher's reputation but, despite warnings from his friends at other publishing houses, he signed a contract for the book to be released under his own name. His associates regarded this move as literary suicide and waited for the inevitable fall out. When *Lolita* was eventually released in September 1955 it was badly produced, poorly translated and full of spelling errors. Although the initial print run of 5,000 copies soon sold out, no newspaper or magazine dared to produce a review, even in France. However, by the end of 1955 the English novelist Graham Greene wrote a piece for the *Sunday Times* describing *Lolita* as 'one of the three best books of 1955'. This finally provoked a response, with the *Sunday Express* calling the book 'filth'. The Home Office reacted by banning *Lolita* and publicly instructing customs officials to seize all copies. It was the sort of publicity of which Nabokov and his publisher could only have dreamed.

The French authorities, who finally realized what was going on, also banned *Lolita*. When restrictions finally eased two years later, Weidenfeld & Nicolson bought the UK rights, although the ensuing scandal ended the political aspirations of one of its partners, Nigel Nicolson, MP, who was forced to step down after his constituency association refused to endorse him for the 1959 general election. He didn't need to be too concerned by this as *Lolita* went on to sell over fifty million copies and was adapted into a film both in 1962 by Stanley Kubrick and again in 1997 by Adrian Lyne. The story has been produced many times as a play, twice as an opera, twice as a ballet and once as a musical, and is listed as number 4 on the Modern Library's list of the 100 best novels of the twentieth century.

It is apparent to me that the possibilities of the aeroplane, which two or three years ago were thought to hold the solution to the problem, have been exhausted and that we must turn elsewhere.

Thomas Edison, American inventor, in 1895

Harry Potter and the Philosopher's Stone by J. K. Rowling
In December 1993 the former teacher, and would-be

author, Joanne Rowling (b. 1965) was living alone in a small rented flat in Edinburgh, Scotland, having run away from the father of her six-month-old daughter in Portugal. All she had was the first three chapters of a story she had thought up whilst travelling on a crowded train to London a few years earlier, which she had begun to write down between teaching jobs in the city of Porto over the previous year. With little else to do, the single mother would take her baby out walking and, as soon as she fell asleep, settle into a café to work on more chapters of the story she had been developing.

In 1995 Rowling completed the first draft of the story she called *Harry Potter and the Philosopher's Stone* and submitted the first three chapters to the Christopher Little Literary Agency, who agreed to represent her. Over the next twelve months the agency received twelve rejections from editors who dismissed the manuscript as 'too long', and nobody, it seemed, was interested in the story. Rowling, by her own admission, felt a failure and admitted this in 2008 when she revealed: 'Had I really succeeded at anything else, I might never have found the determination to succeed in the one area where I truly belonged. I was set free, because my greatest fear had been realized, and I was still alive, and I still had a daughter whom I adored, and I had an old typewriter, and a big idea.'

During this period Rowling received welfare benefits until, in 1996, Barry Cunningham, an editor at Bloomsbury publishers in London, took some chapters home with him at the weekend to do some reading and catch up on his submissions. But it wasn't until Alice Newton, the eight-year-old daughter of chief executive Nigel

Newton, was given the first chapter to read and immediately demanded the next two, telling her father, 'It is so much better than anything else,' that Bloomsbury made an offer for the manuscript of £1,500 plus royalties from sales, if there were any. Rowling was also advised to get a 'proper job' as children's books were notoriously difficult to sell and 'nobody made any money out of them'.

In 1997 Rowling applied for, and received, an £8,000 grant from the Scottish Arts Council, which enabled her to continue writing for a while and in June 1997 an initial print run of only 500 copies was distributed, 300 of which were given to libraries. First edition copies from this initial print run now change hands between collectors for in excess of £40,000. Sales were initially very slow, although between July and September positive reviews started to appear in the local press and then the nationals began to follow suit. However, it was thanks to the publicity gained after *Harry Potter and the Philosopher's Stone* won a National Book Award six months after its publication that the *Harry Potter* juggernaut started to develop into the $15 billion book and film industry that it has become today.

MORE AUTHORS WHO SUFFERED BRUTAL REJECTIONS

In truth, the book publishing industry is no place for the soft-souled, and its history is littered with examples of writers who were told they had terrible ideas before they went on to achieve global success.

When forty-six-year-old former Pan Am air hostess Mary Higgins Clark (b. 1927) submitted the manuscript for her first novel, *Where Are the Children*, in 1975, she was told, 'We find the heroine to be boring.' The book is now in its seventy-fifth print run and became the first of Clark's forty-two bestselling books, earning her in excess of $60 million in the process.

The Wind in the Willows author Kenneth Grahame (1859–1932) was told, 'It is an irresponsible holiday story that will never sell.' Over a century later, and in excess of twenty-five million copies later, the book still sells very nicely, thank you.

'It is so badly written' was the verdict Dan Brown (b. 1964) received for *The Da Vinci Code* before he went on to become the twentieth-biggest-selling author of all time with over 200 million book sales to his credit.

Paulo Coelho (b. 1947) sold 800 copies of *The Alchemist* before finding a new publisher and slightly improving that figure to 75 million.

American dentist Pearl Zane Gray (1872–1939) was told in 1903, 'You have no business being a writer and should give it up.' There are thought to be in excess of 250 million of his books still in print today.

Dr Seuss (1904–91) was once told, 'This is too different from other juveniles on the market to warrant selling this.' Three hundred million sales make Theodor Seuss Geisel the ninth-biggest-selling author of all time.

'Nobody wants to read a book about a seagull,' Richard Bach (b. 1937) was told in a rejection for *Jonathan Livingston Seagull*. So far forty-four million people have, and counting.

Jacqueline Susann (1918–74) was told she was an 'undisciplined, rambling and thoroughly amateurish writer', before *Valley of the Dolls* sold thirty million copies.

Gone with the Wind was rejected thirty-eight times before Margaret Mitchell (1900–49) found a publisher who sold her thirty million copies.

William Golding (1911–93) received a rejection letter informing him that *Lord of the Flies* was 'an absurd and uninteresting fantasy which is rubbish and dull'. Personally, I think the editor had a fair point.

There were rejection letters covering three years under the bed of Meg Cabot (b. 1967), and the bag became so heavy she was unable to lift it before *The Princess Diaries* was finally accepted and sold fifteen million copies.

After being rejected by twenty-five literary agents Audrey Niffenegger (b. 1963) sent an unsolicited copy of her manuscript to a small San Francesco publisher and then held her breath. MacAdam/Cage liked the story, agreed to publish it and *The Time Traveler's Wife* sold seven million copies in thirty-three languages across the world.

Because Garth Stein's (b. 1964) book *The Art of Racing in the Rain* is narrated by a dog, his agent rejected the idea. Stein changed agents and Folio Literary Management immediately sold the rights for more than $1,000,000.

One publisher famously turned down the rights to *The War of the Worlds* by H. G. Wells (1866–1946) with the words, 'It is an endless nightmare. I think the verdict will be, "Oh don't read that horrid little book."' That horrid little book has been in print since 1898 and is regarded as one of the classics.

In 1956 Patrick Dennis (1921–76) became the first author in history to have three books ranked on the *New York Times* bestseller list at the same time. Previously he had submitted his manuscripts to the entire list of American publishers in alphabetical order. His eventual publisher was Vanguard Press.

Alex Haley (1921–92) received 200 rejections over an eight-year period before his book *Roots* was finally published and sold 1.5 million copies in its first eight months.

The Jack London State Historic Park near Glen Ellen in California has a collection of the 600 rejection letters Jack London (1876–1916) received before he sold his first story.

SUPERSTARS WHO WERE TOLD NOT TO GIVE UP THEIR DAY JOBS

The Grand Ole Opry is a country music stage concert founded in 1925 in Nashville, Tennessee. The Opry, which began as a weekly barn dance that was listened to on WSM Radio (famously known as the Legend), began broadcasting nationally in 1939 and is now known as one of the longest running radio programmes in history. Over the years the Opry has provided a stage for country music icons such as Hank Williams, Patsy Cline and the Carter Family, establishing the city of Nashville as the home of the world's country and western genre in the process. More recently stars such as Dolly Parton, the Dixie Chicks and Garth Brookes have found a home upon its stage.

However, on 2 October 1954 a gangly nineteen-year-old hopeful made his first and only appearance, to which the live audience reacted politely, despite finding his brash music and 'snake-hip' gyrations to be 'vulgar and distasteful'. The General Manager at the time, Jim Denny, later told young Elvis Presley (1935–77) that he should return home to Memphis and, 'You ain't going nowhere, son, except back to driving a truck.' Only six months earlier the teenager had failed two auditions to become the vocalist in local bands on the grounds that he couldn't sing. The following month, in November 1954, Presley was offered a contract to perform fifty-two shows at the

Louisiana Hayride where he famously met Colonel Tom Parker, who had other ideas for him.

Within twelve months, and towards the end of the Hayride residency, Presley had not only been voted the year's top artist at the Country Disk Jockey Convention, but Colonel Tom Parker was considering offers of $25,000 each from three record labels for the young star's signature. On 21 November 1955 Parker received an offer from RCA of an unprecedented $40,000, which the pair eagerly accepted, although Presley, still only a minor, had to ask his father to sign the papers. Within another year Elvis Presley would become the most famous singer in the world and, by the time of his death twenty years later, had contributed over 100 singles to the *Billboard* charts, more than any other solo artist in history.

But Elvis is not the only iconic musician who would suffer the effects of rejection and/or misjudgement. In 1967 Jimi Hendrix (1942–70) had reached the top ten of the UK charts three times but had so far failed to make an impression in his native America, other than as a respected session guitarist. One day Mike Nesmith of the Monkees was having dinner in London with Paul McCartney and Eric Clapton when John Lennon arrived. Nesmith later recalled: 'John said, "Sorry I am late but I've got something I want to play to you guys." He had a tape recorder and played Jimi Hendrix's "Hey Joe". Everybody's mouth just dropped open. John said, "Isn't this wonderful?"' A few weeks later Micky Dolenz of the Monkees met Hendrix at the Monterey Pop Festival and suggested to his band's producers that they invite the Jimi Hendrix Experience to support them on their American Tour.

As ridiculous as the idea seems now, Hendrix was persuaded by his manager, Chas Chandler, to accept the offer that would expose his music to hundreds of thousands of young record-buying American kids. It was a decision that would backfire spectacularly because, despite the Monkees loving Hendrix and his music (they would sit on the floor and listen to the sound checks), their fans jeered and booed. Micky Dolenz later recalled, 'Jimi would amble onto the stage, crank up the amps and launch into "Purple Haze" and the kids in the audience instantly drowned him out with, "We want Davy." God it was embarrassing.' Within seven days Hendrix parted with the tour. However, the often repeated story that the great Jimi Hendrix was dropped from the Monkees tour is not true. At that time 'Purple Haze' was making an impression in the US charts and Hendrix was in demand from his own fans so he was allowed to break the touring agreement amicably.

> Heavier-than-air flying machines are impossible.
> Lord Kelvin, British mathematician and physicist,
> President of the British Royal Society, in 1895

In 1963 the powerful music agent Eric Easton thought that a new, hot group in London had some potential on the busy live music circuit. 'But that singer will have to go,' he told their manager, 'the BBC won't like him.' And the BBC certainly didn't like Mick Jagger very much in the early days of the Rolling Stones' career.

In 1944 Marilyn Monroe (1926–62) was spotted by a fashion photographer who encouraged her to apply to the Blue Book modelling agency in the hope of securing an agreement with the firm. One of the directors,

Emmeline Snively, explained to the young hopeful that they were only looking for models with blonde hair and advised Marilyn, a brunette, to 'go and learn how to be a secretary or get married'. Instead, Marilyn dyed her hair golden blonde, returned the following week, was offered a contract and became the Blue Book's most successful client. And Marilyn wasn't to be the last who was discriminated against. The fabulous Salma Hayek (b. 1966) was told that she would never make it as an actress because once people heard her speak, they would think of their maids.

By 1964 Ronald Reagan (1911–2004) had spent most of his career as a Hollywood actor, albeit mainly in what were known as 'B' movies. In 1964 he auditioned, and was once again rejected, for the role of a presidential candidate. He was turned down because studio executives declared that Reagan 'did not have the presidential look'. Fourteen years later he accepted the Republican nomination for the role of the real president and was duly elected in 1980.

> The cinema is little more than a fad. It's canned drama. What audiences really want to see is flesh and blood on the stage.
> Actor, producer and comedian Charlie Chaplin, 1916

MONOPOLY
Board games have been popular throughout the world for over 5,000 years and have often been associated with learning and arithmetic. In 1903 Lizzie Magie (1866–1948), an American follower of economist Henry George (1839–97), realized that the public did not understand

how land and property rental served only to increase the wealth of the rich whilst their tenants would remain in poverty. The Georgist economic philosophy included the idea that everything found in nature belonged equally to everybody. Henry George nobly argued that a tax on land values would eventually reduce the inequality between the haves and the have-nots, although he appears not to have considered the idea that the landowners would simply pass on this extra cost to their tenants and make their lives even less bearable.

However, Lizzie Magie felt that this Georgist ideology would be better understood by the minions if she could demonstrate the complicated relationship between land-owner and tenant by inventing a board game that could teach them how it actually worked in practice. In 1904 Magie was granted a patent for a board game she called 'The Landlord's Game', which was innovative for several reasons: the first being that there was no 'end square' for gamers to reach as there is in backgammon and other board games, and the second was that the object, the way to win, was mass ownership. It was the capitalist principle of acquiring property and land in order to amass great wealth at the expense of others. From a finite bank (the economy) some would grow rich and charge others rents, which would eventually lead to their bankruptcy and the loss of the game. Just as in real life.

Initially popular with other Georgists, The Landlord's Game was not manufactured until 1906 and, unsurprisingly, failed to prove popular. Soon the Parker Brothers, successful toy and game manufactures with over twenty-five years of experience, were offered the rights but dismissed the game as 'too complicated'. The Parker Brothers had been responsible for Rook in 1906, which had become America's most successful card game, and they thought they understood the market place. Although The Landlord's Game continued to be used as a teaching tool in economics classes and its limited popularity spread through word of mouth, particularly within the Quaker communities who used hand-made boards, there was to be no national distribution.

> I'm just glad it will be Clark Gable who is falling on his face and not Gary Cooper.
>> Gary Cooper explaining why he turned down the lead role in *Gone with the Wind* (1939)

In 1929 salesman Charles B. Darrow (1889–1967) lost his job as a result of the great stock-market crash of that year and began to take up a series of odd jobs around his home neighbourhood in Philadelphia. It was then that he noticed his friends and neighbours playing a homemade board game, which involved buying and selling property and was almost certainly a direct descendant of Lizzie Magie's The Landlord Game, and decided to come up with his own version. The one he had been playing at home had been brought back to Philadelphia from Atlanta by his friend Charles Todd, who explained that the street names and other locations on the board were all to be found in Atlanta. The game Darrow developed was identical to the Atlanta City version, although he did introduce the icons for the electrical and water works and the stations that would later help to make Monopoly the most famous board game in the world.

But he still had work to do. In 1933, when he approached the same Parker Brothers as Lizzie Magie had thirty years earlier, his game was also rejected as 'too

126

complicated'. The world was in recession and the Parkers were not planning on introducing any expensive new prototypes. Undeterred, Darrow arranged for a friend of his who owned a printing business to produce 5,000 copies after a department store in Philadelphia had agreed to give his game a trial run in 1935. Everybody involved was then amazed that, during a time of recession, people loved playing a game that could bring them huge fortunes, albeit fantasy fortunes. Monopoly immediately sold its entire print run and soon orders were pouring in. The Parker Brothers noticed this enthusiasm and approached Darrow with a firm offer to buy his game, which then sold an estimated 200 million copies.

As expected, Lizzie Magie also noticed the enormous success of Monopoly, although, as her patent had expired in 1921, there was little she could do about it. Despite her criticisms of the Parker Brothers in the national press, they did agree to publish two more of her games in 1937 and The Landlord's Game made a third appearance in 1939. By then, however, Monopoly dominated the market. Charles B. Darrow had become a millionaire and went down in history as the man who invented the most popular board game of all time.

How Wrong Can You Be?
Arctic expert Bernt Balchen made a claim in the *Christian Science Monitor*, on 8 June 1972, that a general warming trend over the North Pole was melting the polar ice cap and may produce an ice-free Arctic Ocean by the year 2000.

THE PET ROCK

In April 1975 a struggling freelance advertising copy-writer was having an after-work beer with some friends in Los Gatos, California. As the conversation moved from idle moaning about a lack of business to complaining about their pets, thirty-five-year-old Gary Dahl informed the group that he found cats, dogs, fish and birds all to be a 'pain in the arse' as they took too much looking after, made a mess, were expensive and generally mis-behaved. Instead, he announced, he had a pet rock that was cheap, easy to look after and gave him hours of enter-tainment. It was, he declared, the perfect companion. The group then started discussing all the possible advantages of a pet rock, which became more and more inventive as the drinks flowed, and Dahl went home that evening with a plan.

Over the next two weeks Gary Dahl gathered all the ideas together and began compiling the 'Pet Rock Train-ing Manual', which was a detailed guide to owning and caring for a rock. Included in the manual were tips on how to train it to do tricks such as 'roll over' (best prac-tised on a steep hill) and 'play dead' (they prefer to be alone), and how to take it for walks (in your pocket). The idea was to release the manual as a novelty gift book but Dahl went a step further and decided to include the actual pet. At his local builders' supply yard the writer found Rosarita beach stone: rounded grey pebbles of uni-form size that, most importantly, he could buy for a penny each. He then set about creating a small cardboard box complete with holes and a drinking straw and his new pet was ready for market.

By the end of October over half a million Pet Rocks had been sold for $3.95 each, with Gary earning a dollar in profit from each sale. He had become a wealthy man in only two months, and the Pet Rock craze was still gripping the country, with rival entrepreneurs quickly creating Pet Rock Obedience Lessons and offering Pet Rock Burial Services. In the run up to Christmas 1975 it is estimated that over two million Pet Rocks were sold, earning Gary Dahl just over $2 million from his venture. The Inland Revenue Service, no doubt, was on high alert. Before long, as 1975 became 1976, the Pet Rock craze slowed down to a trickle and the fad was over, but Gary Dahl retired from his job in advertising a very happy man.

Gary soon opened a bar in his home town called Carrie Nation's, after the famous Temperance Movement campaigner, and later released small boxes of Red China Dirt along with a marketing campaign that promised to 'smuggle mainland China into the United States one cubic centimetre at a time'. Unfortunately, his new idea failed to repeat the success of his Pet Rock, and so Gary formed his own advertising agency and has since created thousands of radio and television adverts that have earned the agency a variety of awards along the way. The Pet Rock, eh? We all hope to have one of those ideas in us, don't we?

TV will never be a serious competitor for the radio because people must sit and keep their eyes glued on a screen. The average American family hasn't the time for it.
New York Times, 1939

BILLY BOB'S TEETH
In 1994 Missouri State University graduate Jonah White was living in a cave, where he had been contemplating 'get rich' schemes for over a year, when his old football

coach, Jess Branch, left a message for him at his parents' house. White had been a champion footballer and the message included an invitation to return and speak to his old Bears team who were on a serious losing streak. Jonah White was well remembered at his old college and was received with enthusiasm from the new, younger players in the team. Whilst there, White noticed another student, Richard Bailey, talking confidently to a group of girls and yet he had the ugliest, yellow/black buck teeth White had ever seen in his life. White later recalled, 'But the guy oozed so much confidence and was like a body-builder. It was hard to believe he would take care of his body like that, but not his teeth.'

Bailey later recognized White as an old school football star, and when he approached him to shake hands White was surprised to find he had perfectly straight, clean white teeth. The dental student then pulled out an ugly pair of dentures and cried, 'Hey, how do you like my Billy Bob Teeth?' White and Bailey immediately formed a friendship and within three weeks Jonah White had sold off his personal possessions, which amounted to a Springfield .45 handgun, and used the money to incorporate Billy Bob Products. For the next three years the pair travelled around bars and shopping malls selling Billy Bob Teeth and increasing their range of products to include Billy Bob Pacifiers to 'make sure your kid doesn't get abducted'.

Along the way the pair were repeatedly told they were fools and should be looking for regular work, despite increasing sales of up to 30 per cent per month. White remembers this period well and later revealed: 'I had no doubt it would be huge. Still, I'll admit I never knew it would be this big. My goal was to sell a million pieces.

130

Ninety-nine percent of people told me I was a fool and I'd be out of business in no time.'

Today the ugly hillbilly teeth, baby pacifiers and over 300 other products have sold in excess of 40 million units and are exported to 95 per cent of the countries in the world. White himself is said to have a personal fortune of $50 million and lives with his family in an 8,000-square-foot mansion on a 900-acre estate. Not bad for a lad who grew up in a log cabin, lived in a cave and thought bad teeth would be a great business idea.

THE GRAND CANYON

The first people known to have lived in the area of the Grand Canyon were the ancient Puebloans whose descendants occupied the land from around 1200 BC until the American settlement of the West during the nineteenth century. The Pueblos first encountered Europeans in AD 1540 when the Spanish conquistadors arrived and explored the region. (They lived in villages the Spanish called '*pueblos*', which translates as 'towns'.) At that time the native people called the vast canyon '*ongtupqa*' and considered it a holy site. For centuries they had made regular pilgrimages to it whilst others settled there and lived within the many caves in the area. However, García López de Cárdenas, the first European to visit the area, and who was searching for the fabled Seven Cities of Cibola, which were rumoured to be full of 'limitless gold', found nothing of any use or value. His group attempted to descend into the valley to collect fresh water but failed to make it even a third of the way down before they were forced back because they ran out of supplies. It is generally believed that their Pueblo guides, who must have known the safe routes down, were reluctant to lead them to the sacred river. Cárdenas and his men soon left and no Europeans returned to Ongtupqa for another 200 years until, in 1776, two Spanish priests and a small group of soldiers set out to find a route between Sante Fe and California.

Father Silvestre Vélez de Escalante and his companion Father Francisco Atanasio Domínguez explored southern Utah before reaching the Grand Canyon and navigating the north rim to find a crossing that would become known as the 'Crossing of the Fathers'. They also did not stay long but did describe the vast canyon in their expedition reports. They were followed in the same year by Father Francisco Garcés, who was a missionary. He spent a week in the area attempting to convert the natives to Christianity, without success, before leaving. He later described the canyon as 'profound'. Next along was a group of East Coast trappers led by James Ohio Pattie, who arrived in 1826 and who again almost as soon as they had arrived, having also discovered nothing of any value. During the 1850s a Mormon missionary by the name of Jacob Hamblin (1819–96) was sent by Brigham Young (1801–77), a church leader who commissioned a party to establish river crossings and build 'good relations' with the native Americans and the few European settlers who had already blazed a trail in the general direction.

> How, sir, would you make a ship sail against the wind and currents by lighting a bonfire under her deck? I pray you excuse me. I have no time to listen to such nonsense.
> French Emperor Napoleon Bonaparte on being offered the world's first steamboat design in 1800 by American inventor Robert Fulton

Hamblin discovered the 'Crossing of the Fathers' and helped another missionary, John Doyle Lee, establish Lee's Ferry, which doubled as a Mormon outpost and operated for sixty years before being replaced by a bridge. (Lee's Ferry is now famous as a fishing and boat launching point that includes white-water rafting trips through the Grand Canyon.) Hamblin later acted as an advisor and guide to John Wesley Powell (1834–1902), an American soldier, geologist and explorer who famously led the Powell Geographic Expedition of 1869 down the Colorado River and who was the first to pass through the canyon by boat, encountering the dangerous white-water rapids along the way. Hamblin, who had been in the area for fifteen years and had become acquainted with the locals, served as a diplomat between the natives and the explorers, and ensured Powell and his team's safety. John Wesley Powell later became the first man to describe the 'big canyon', as it had become known to English-speaking locals, as the Grand Canyon.

In 1857 Edward Beale led an expedition to survey a wagon trail between Fort Defiance in Arizona and the Colorado River. On 19 September one of his men, May Humphreys Stacy, recorded in his journal that they had discovered 'a wonderful canyon of four thousand feet deep. Everyone has admitted that he never before saw anything to match or equal this astonishing natural curiosity.' At the same time the US War Department had commissioned Lieutenant Joseph Ives to take a party of men to carry out an 'up river' navigation of the Colorado River from the Gulf of California. Aboard a steamboat called the *Explorer*, the group endured two months in difficult conditions before their vessel hit a rock and had to be abandoned. Ives then led his men east along what is now known as Diamond Creek (it was completely unknown at the time) and into the canyon. He later noted, in his report of 1861 to the Senate, that only one or two trappers had previously seen the canyon. Ives, just like the other Europeans before him, also found nothing of any interest or value at the canyon and wrote in his report that 'ours has been the first, and doubtless to be the last, to visit this profitless locality'.

Until then, of course, most explorers had been searching for gold, silver or other mineral wealth. They had failed to anticipate either the strategic value of the Grand Canyon's position on the Colorado River or the impact of tourism as the East Coast dwellers, and more European settlers, would surge west over the next twenty-five years. In fact, there was so little interest in the canyon that when explorers and trailblazers reached the rim and peered down into it, they regarded it as nothing more than an obstacle. It was an impassable chasm. Lifeless, pointless and undesirable. A place where nobody could ever settle and that nobody would ever want to visit intentionally.

And then, in 1889, Frank M. Brown proposed a railroad that would run alongside the Colorado River and would be used to transport coal and other supplies vital to the growing number of settlers moving west. Along with his chief engineer, Robert Brewster Stanton, and a small team, he began exploring the canyon. Unfortunately, with badly built boats and no life jackets, Brown was drowned in an accident near Marble Canyon, further reinforcing the idea that there was nothing for which it was worth travelling to the area. That was until US President Theodore Roosevelt (1858–1919) visited the canyon in 1903. Roosevelt, a keen and rugged out-doorsman and pioneer conservationist, fell in love with what he discovered and on 28 November 1906 he established the Grand Canyon Game Reserve. Roosevelt also encouraged restrictions on livestock and grazing, and chased away predators such as wolves, eagles and mountain lions, making the area safe for visitors.

He then incorporated neighbouring forests and woodland and, on 11 January 1908, declared the preserve a US National Monument. But Roosevelt had plenty of opposition in the shape of miners and claim-holders who successfully opposed his initiatives for the next eleven years until, on 26 February 1919, President Woodrow Wilson (1856–1924) signed an Act of Congress and the Grand Canyon National Park was finally established. Since then, the region that had been, only fifty years earlier, dismissed as a 'profitless locality that nobody would ever want to visit' has become one of the world's leading tourist destinations, attracting more than five million visitors every year. Indeed, this profitless locality now contributes an estimated $500,000,000 (half a billion) to the economy each year.

How Wrong Can You Be?
Dr David Viner, a senior research scientist at the Climatic Research Unit (CRU) of the University of East Anglia, claimed in an interview with the *Independent* on 20 March 2000 that 'children just aren't going to know what snow is. Winter snowfall will become a very rare and exciting event.'

135

THE YO-YO

The simple yo-yo has been a popular toy for over 2,500 years. In fact, it is widely regarded as the second oldest toy of them all (the oldest being the doll). In ancient Greece the yo-yo was made from terracotta, metal and wood, and the two halves were usually decorated with images of the gods. As Greek children became adults they would usually place their favourite yo-yo onto the family altar as part of their rite of passage, in a show of respect to those gods and to symbolize their maturity.

In the Philippines the yo-yo was recognized as an effective weapon, with versions that were studded with spikes and had razor-sharp edges. They also had twenty-foot long cords and were used against both enemy and prey for over four centuries. During the nineteenth century in Europe the English had a similar toy called the bandalore and the French played with their '*incroyables*'. But it was as recently as 1920 that the world came to know the word yo-yo, after a Philippine immigrant by the name of Pedro Flores began manufacturing them at his small toy factory in California. He marketed them as yo-yos, the Tagalog (native Philippine language) word *yoyo* meaning 'to come back', although very few people understood them or took much of an interest in what was essentially a primitive weapon.

That was until Donald F. Duncan Sr (1892–1971), an American inventor, entrepreneur and founder of the Duncan Toys Company, happened to see one during the 1920s and bought the rights to the design from Flores in 1929, quickly trademarking the yo-yo brand in the process. Duncan then modified it by replacing the cord with a slip-string that ensured the yo-yo returned to the hand (not in my case it doesn't) and then invented easy tricks that children loved to try out. Duncan then did a deal with William Randolph Hearst (1863–1951), the newspaper tycoon, to receive free advertising and editorial for his yo-yo in return for running national competitions whereby the entrants had to deliver a certain number of newspaper subscriptions as their entry fee.

Soon, satellite competitions were being organized all over America and youngsters were pleading with relatives and neighbours to subscribe to one of Hearst's tabloids so they would be able to enter. The strategy worked and folk of all ages were soon practising their yo-yo skills wherever they went. Before long, Duncan's factory was producing over 3,500 yo-yos every hour and, after one particular media campaign, sold an incredible three million units in one month alone during 1931. However, as the joke goes, yo-yo sales were notoriously unpredictable and, well, a little up and down. Generally speaking, however, sales continued to increase and reached a monumental peak in 1962 when the Duncan yo-yo sold forty-five million units during that year alone. Even so, Duncan's bold advertising and marketing strategy meant the company spent more than it made. Three years later Duncan lost the yo-yo trademark when a court ruled that the word had become part of the common language and could no longer be owned by a brand. Soon afterwards Duncan was filing for bankruptcy and sold his interests in the yo-yo to the Flambeau Plastic Company. A victim of his own success, if ever there was one.

Happily, that is not the end of the Donald Duncan story because in 1936, using some of the early profits from his yo-yo sales, he began to experiment with designs for a parking meter, something that was first successfully

used in Oklahoma the previous year. Ever the entrepreneur, Duncan marketed his designs with such flair and
originality that by the time he sold the Duncan Parking
Meter Company in 1959 it was manufacturing 80 per cent
of all meters that were being installed around the world.
So the next time you are parking your car in Paris,
London or any of the major cities, check the meter to see
if it bears the Duncan logo. If it does then you have an
old Filipino hunting tool to thank for your parking fine.

BLOCKBUSTER FILMS INITIALLY REJECTED AS TERRIBLE IDEAS

Star Wars

Hard as it is to imagine, when young film-maker George
Lucas (b. 1944), fresh from his success with *American
Graffiti* (1973), approached Universal Pictures and
United Artists with a two-page treatment for a new film
idea, he was flatly rejected. Mind you, the film was to
be called 'Adventures of Luke Starkiller as taken from
the Journal of the Whills, Saga I: The Star Wars'. And the
proposal outline began: 'This is the story of Mace Windy,
a revered Jedi-Bendu of Ophuchi as related to us by
C. J. Thorpe, padawaan learner to the famed Jedi.' Who,
in their right minds, would turn that down? Legend has
it Lucas himself wasn't even that keen on the idea and
instead wanted to remake the 1930s science-fiction series
Flash Gordon, but somebody else beat him to the rights.

And so, with a 200-page script that even his close
friends admitted they didn't understand, Lucas hawked
his film around the major Hollywood studios, and eventually Twentieth Century Fox decided to risk $8.5million, a
modest sum in the world of film-making, even in those
days. George Lucas then assembled a team of actors and

special-effects experts, flew to England and on 22 March 1976 began filming. Within months half of the budget had been spent and the team had only three usable scenes in the can. Sound effects, such as they were in those days, included dropping a refrigerator onto the ground in order to simulate the Death Star explosion. Soon the crew were openly mocking the idea, refusing to work overtime and taking unauthorized breaks, which meant production costs soared.

Once the film had been completed, Twentieth Century Fox took one look at it and realized they had an expensive flop on their hands. There simply was no appetite for that kind of movie and cinema chains refused to show it. Fox executives responded by insisting cinemas had to screen *Star Wars* (Lucas had been persuaded to shorten the name) if they were going to have the eagerly awaited *The Other Side of Midnight* that was scheduled for later in the year and widely predicted to be a box-office smash. Some of the smaller cinema managers relented and *Star Wars* eventually opened in only thirty-nine locations. All Twentieth Century Fox could hope for was to at least make some money back and then wait for their 'guaranteed hit' later in the year. Even at this point, so the story goes, director George Lucas was still convinced the film was a failure, and when he showed it to his friends they all agreed with him. All of them, that is, apart from a certain Steven Spielberg (b. 1946), director of the recent smash hit *Jaws*, who was convinced *Star Wars* would do well at the box office.

Lucas still didn't agree and didn't even bother attending the premiere; he and Spielberg flew off on holiday instead. Nobody at all predicted that the opening night word-of-mouth enthusiasm would lead to queues around the block at all thirty-nine cinemas the next day. This would be followed by the sort of international hype of which no artist would ever dare to dream. *Star Wars* had soon made over $750,000,000 at the box office, and became a multi-billion dollar industry in terms of merchandise and sequels. A multi-billion dollar industry that nobody, including its creator, even contemplated.

Pirates of the Caribbean

In 1986 controversial film director Roman Polanski (b. 1933) managed to persuade some poor soul to give him a $40 million budget to make a pirate film starring Walter Matthau and called, imaginatively enough, *Pirates*. The film returned around $8 million at the box office before disappearing without a trace. A decade later and Hollywood had another go. And why not, pirates are always popular, right? This time *Cutthroat Island* cost $100 million to make and returned only $10 million in ticket sales. Its single accolade was an acknowledgement from *The Guinness Book of Records* as 'the biggest box office flop of all time'. Even the Muppets, with their successful global brand, only managed to return $34 million for their *Muppet Treasure Island*. Although they did make a profit thanks to their low actors-wages overhead. And then there was the Disney produced *Treasure Planet* in 2002, which managed to make a loss of only $30 million. It could have been much worse.

With a recent track record like that, film producers could have been forgiven for wanting to avoid pirate films for a few decades, and one can only wonder what it was, or who it was, at Disney that managed to persuade them to approach the bean counters and ask for the $140 million they needed for yet another pirate film the following year. Everybody was against it and nobody thought it was a good idea. Even the proposed choice for leading actor was considered risky, as two of Johnny Depp's previous films, *The Man Who Cried* and *Before Night Falls*, had failed to return anything like the proposed budget for *Pirates of the Caribbean* at the box office. In fact, they had grossed around only $10 million between them. It seemed to some as if Disney were intentionally trying to lose money on pirate films. Instead, *Pirates of the Caribbean: The Curse of the Black Pearl* made over $650 million in ticket sales and was the start of one of the biggest-selling film series of all time. There must have been executives at Disney who were not surprised, but everybody else was. Mind you, the Disney empire was entirely built on these types of risk (see '*Snow White and the Seven Dwarfs*' and 'Glorious Failures').

Raiders of the Lost Ark

After their combined, and massive, success during the 1970s with *Jaws* and *Star Wars*, young film-makers George Lucas and Steven Spielberg were perfectly entitled to expect Hollywood studios to be queuing up for a joint venture they had dreamt up between them. On paper, *Raiders of the Lost Ark* had everything going for it. A decent script with a mix of action and humour, two directors with box office appeal and a strong cast. But the studio executives, it seemed, weren't particularly interested in a story of a relic-hunting archaeologist, set shortly before the Second World War. One of the problems may have been that Spielberg's previous film, *1941*, set during the Second World War, had panned. *Raiders* was turned down by virtually every major Hollywood studio and dismissed as 'too expensive' to be worth making.

141

Eventually Paramount stepped in and, despite Spielberg and Lucas's growing reputations, only agreed to finance the project to the tune of $18 million, a paltry sum for a pair of film makers who had just returned $1.3 billion from *Jaws* and *Star Wars*, two of the biggest films ever made, using a combined budget of only $20 million. To save costs Spielberg made the film at Elstree in England, using the minimum amount of takes for each scene. The result, the film that was turned down by every major Hollywood producer, is now considered to be a master-piece of film-making and grossed the best part of $400 million at the box office.

Snow White and the Seven Dwarfs

'I have an idea,' a wise man once said. 'How about we make a full-length animated cartoon out of one of those old German fairy tales? The ones that the Brothers Grimm made popular around 120 years ago? We could raise a budget of far more than our whole company is worth, spend it all and see what happens. The only time it has ever been tried before, it was a complete failure, a flop, it bombed. Everybody hates full-length cartoons. So let's do it.' How they must have all laughed. In truth, we have no idea what was said at that meeting of Disney executives in 1936 but we can assume that not everybody laughed. Because on 21 December 1937, *Snow White and the Seven Dwarfs* premiered at the Carthay Circle Theatre on San Vicente Boulevard, Hollywood.

After blowing the best part of $1.5 million on the production (and remember, this was a time when a six-bedroom Hollywood mansion would cost around $6,000), the film was already being dismissed as 'Disney's folly', and both Walt Disney's brother and wife had tried to talk him out of making it. Walt Disney (1901–66) himself was so out of his comfort zone that he had no idea how to distribute the film or how much to charge for tickets. Eventually Charlie Chaplin showed him the reports from his own successful films to use as a reference. By this time Disney might just as well have gone down to Las Vegas and put everything he had on black or red. A national release followed on 4 February 1938 and *Snow White* earned Walt Disney $8million during the first few weeks alone. It quickly became the highest-grossing sound film of its time and grew to be so popular that whenever the Disney Corporation needed money, they re-released *Snow White*. As they did in 1944, 1952, 1958, 1967, 1975, 1983 and 1987. When it was released again in 1993 the fifty-six-year-old film went straight into the top ten of the box office listings. To date, *Snow White* has earned Disney nearly half a billion dollars on their $1.5 million investment and, during its first run at the cinemas, the film sold over 109 million tickets which, at today's prices, would make it one of the most financially successful films in the history of cinema. And that is to say nothing of the animated film industry it spawned.

> I must confess that my imagination refuses to see any sort of submarine doing anything but suffocating its crew and floundering at sea.
> British science-fiction writer H. G. Wells

Back to the Future

In 1981 scriptwriter and producer Bob Gale (b. 1951) had a script for a film he thought would be perfect for its time. Teen comedy was so popular that Gale was convinced he had a winner on his hands. Columbia, however, thought the script was too family friendly to be a success and rejected the idea. Gale later remembered: 'They thought it was a really nice, cute, warm film but not sexual enough. They suggested that we take it to Disney, but we decided to see if any of the other major studios wanted a piece of us.' The problem for Gale was that none of them did. Disney, unsurprisingly, decided that a plot involving a kid that travels back in time only for his own mother to fall romantically in love with him was inappropriate for the Disney family brand.

The film looked set for the shelf until its director, Robert Zemeckis (b. 1952), had a major hit with *Romancing the Stone* in 1984 and the project started coming together with Universal Studios. But even the early stages of filming were slow and after only four weeks Zemeckis decided it would have to be re-cast. It was a decision that would add $3 million to the production costs but, in securing the services of Michael J. Fox, the director felt he was ready to start filming again after four years of rejection. On its release in 1985, *Back to the Future* returned nearly $400 million from a budget of $19 million and became one of the most successful films of the decade, earning all sorts of nominations and awards. It even got a namecheck during US President Ronald Reagan's 1986 State of the Union address.

Home Alone

For the want of a nail . . . the kingdom was lost. Or so the saying goes. Something similar happened during the filming of *Home Alone*, which had been given a production budget of $14 million. During the later stages of filming, director Chris Columbus (b. 1958) went back to Warner Brothers with his cap in hand. He requested a further

$3 million to finish the project in the way he wanted and was sent packing by studio executives. They refused to invest another cent. Columbus then called his associates at Twentieth Century Fox and asked, 'Would you like the picture?' Apparently it took Fox only twenty minutes to make a positive decision and a deal was thrashed out with Warner Brothers for them to take over the project.

Ironically, on the film's opening weekend it earned exactly $17million and then became the number one film at the box office for twelve consecutive weeks. By the end of its run in cinemas *Home Alone* had taken nearly half a billion dollars at the box office and had become the third-highest grossing film of all time. Including the sequels, the franchise has topped a billion dollars. And that was probably the worst $3million saving ever made by Warner Brothers or anybody else.

Pulp Fiction

Columbia Tristar considered the treatment for *Pulp Fiction* to be 'the worst thing ever written. It makes no sense. Someone's dead and then they are alive. It's too long, too violent and un-filmable.' At least that's how co-writer Roger Avary (b. 1965) remembered the conversation. When Columbia boss Mike Medavoy concluded that it was 'too demented to risk', Avary thought, 'Well that's that then.' The proposal was immediately added to the turnaround, which meant it could be sold off to another studio. However, the 1992 success of *Reservoir Dogs* by the same director, Quentin Tarantino (b. 1963), encouraged Miramax's Harvey Weinstein to adopt the project. Weinstein was not troubled by the film's portrayal of heroin use and gratuitous violence, or its cast of fading stars, and *Pulp Fiction* became the studio's first signing after being taken over by Disney.

After the film's release on 14 October 1994, Tarantino received an Academy Award for Best Screenplay, which he shared with Roger Avary, and he was also nominated for Best Director. The film received five other nominations and won the Palme d'Or at the Cannes Film Festival earlier in the year. It was also instrumental in reviving the careers of the lead actors, Samuel L. Jackson, Bruce Willis and John Travolta, and, finally, there was the small matter of earning Miramax a comfortable $220 million in return for their modest investment of $8 million.

GLORIOUS FAILURES

Thomas Edison (1847–1931) was told by his teachers that he was too stupid to learn anything. He was dismissed from his first two jobs for being unproductive and famously made a thousand failed attempts to produce an electric light bulb. He was later asked how it felt to fail a thousand times and replied, 'I didn't. The light bulb was invented in one thousand steps.'

Sigmund Freud (1856–1939) was booed and jeered from the stage when he first presented his theories to the European scientific community. And so he returned to his research and emerged later to become known as the founding father of psychoanalysis.

Winston Churchill (1874–1965) failed to achieve grade passes at his early school and later, when he entered the prestigious Harrow School, he was placed in the lowest stream of the lowest class. He then twice failed the entrance exam to the Royal Military Academy at Sandhurst and when he did finally pass he was placed in the cavalry instead of the infantry because he would not need to know any maths. His career as a war correspondent suffered because of his speech impediment, or lisp, and he was defeated in his first attempt to be elected for Parliament. He was soon deselected and had to find another seat to fight for in 1906. Later, once he had been elected, he was sacked from the Admiralty for proposing the disastrous Gallipoli campaign of 1915. From 1931 Churchill found himself in the political wilderness and virtually out of politics altogether, but, just as he was contemplating retirement, the rise of fascism and communism in Europe brought him back into focus. Prime minister for the first time at the age of sixty-two, Churchill then went on to be considered one of the greatest Britons of all time. He later wrote, 'Never give in, never give in, never, never, never. Nothing, great or small, large or petty. Never give up.'

Abraham Lincoln (1809–65) had an informal, rural education that amounted to around one year's worth of classes from several disinterested, unqualified teachers. His first job, after leaving home at twenty-two, was to navigate a raft of merchandise down the Mississippi River from New Salem to New Orleans. He had to walk back. He joined the Illinois Militia as a captain and returned from the Black Hawk War as a humble private. He studied law but lacked the temperament to be a lawyer and, turning to politics, he finished eighth out of thirteen at his first election. He was defeated again in an attempt to be nominated for Congress and had his bid to become Commissioner for the General Land Office rejected. In 1854 he lost a senatorial election and two years later, after helping to reform the Republican Party, lost the election to become its candidate for Vice President. In 1858 he lost another senatorial election and wrote to a friend, 'I am now the most miserable man living. If what I feel now were equally distributed to the whole human family, there would not be one cheerful face on earth.' In 1860 he was elected President of the United States of America and war almost immediately broke out. Once that was over, with Lincoln prevailing, his wife described him as 'happy for the first time'. Weeks later, he was shot in the head.

Socrates (469–399 BC) is credited now as the founder of Western philosophy. In his day he was branded a corrupter of youth. He was portrayed during his lifetime as a clown who taught his students deception, and his ideas and theories frequently clashed with the perceived wisdom of the time. He appeared to even be a critic of democracy itself and dismissed those who were considered to be wise men as 'knowing very little'. He publicly criticized the prominent members of Athenian society, made them appear foolish and accused them of corruption. In response he was arrested, and convicted, for corrupting the minds of Athenians and not believing in the gods of the state. He was asked to propose his own punishment and he recommended a wage to be paid by the government and free dinners for life. In response he was sentenced to death by poisoning. A sentence that he inflicted upon himself by drinking hemlock.

Henry Ford (1863–1947) had a modest education and, in truth, didn't really invent the motor car, and he certainly did not invent the engine or its fuel. In fact, it is hard to believe he was already forty years old when he established the company that would, in his own words, invent the modern age. Ford, a naturally curious boy, started work at the Edison Illuminating Company in 1891 and displayed such enterprise that he was promoted to Chief Engineer within two years. In this position he was able to devote some time to his great passion, engines and automobiles, which he had first learned of during a trip to Detroit. Ford was fascinated the Duryea brothers' first attempts to create a petrol-powered car in early 1896 and resolved to return to his workshop and build his own version. His first effort was the Ford Quadricycle, which was, essentially, two bicycles side by side and powered by a crude engine. On the night of 4 June, Ford carried out his first successful test run, but it would be three years before he could find a backer for his venture and, together, they founded the Detroit Motor Company in 1899.

Sadly, although Ford knew how to build a car, he wasn't able to build enough of them to make a profit, so his investors lost faith and the business closed down after only a year. But Ford wasn't discouraged and formed a new plan to build a racing car. He saw this as the only way to promote his company nationally, and within two years he had successfully raced his first motor. This attracted new investors and the Henry Ford Company was formed in November 1901. However, after a dispute with his fellow directors Ford resigned in 1902 and his former colleagues formed the Cadillac Automobile Company instead. Ford responded by arranging a new partnership (Ford & Malcomson Ltd) with Alexander Malcomson, a coal dealer, and with renewed energy set about building cars. However, the company soon ran into trouble and was in debt to a machine company owned by John and Horace Dodge to the tune of $160,000. The Dodges demanded payment but, with little chance of recovery, accepted shares in his company instead, which was then renamed the Ford Automobile Company in 1903. In July of that year, after seven years of failure, Ford sold his first car to a dentist from Chicago, and within the first year had achieved sales of over 500 more and was already developing the 'Model B'.

Sales remained modest over the following five years, until 1908, when Ford succeeded in making a car that was cheap enough for the mass market, which he called his 'Model T'. It would be another five years until Henry Ford solved his limited production problems by inventing what he called his 'production line', and thousands of Model Ts began leaving the factory to satisfy growing demand. Within a year the Model T had reached sales of over 250,000, and by 1918 half of all cars in America were Fords. In later life he reflected upon his early efforts and concluded that 'failure affords you the opportunity to begin again, more intelligently'. Ford may not have invented the car or the engine, but he did invent an industry that would lead to road building, service stations, fast food, petroleum and the traffic jam. By the time of his death Henry Ford was probably the most influential man of the twentieth century.

Walt Disney (1901–66) was thought to be destined for failure from the very beginning. An average student, he only showed an interest in art, and later vaudeville theatre and early motion films. He was fired from his first job by a newspaper editor who thought his drawings lacked imagination, and he had to work in banking before he managed to find employment as a cartoonist, but the studio soon failed and closed down. At the age of nineteen he and a friend formed their own art studio, which was also forced to close down shortly afterwards. Two years later he tried again and managed to secure a deal with a New York distribution company that promised to promote his work, but would only pay for it after six

151

months. At one point he was so poor he was eating dog food in order to survive and had to, once again, give up and find paid employment. When he finally managed to create his own successful cartoon character in 1926 (Oswald the Rabbit), he attempted to renegotiate his deal with the distributor, Universal Studios, for a better rate of pay, only to be informed that he had signed away his copyright for the character and the studio hired other artists to continue the series.

The following year he created Mickey Mouse and was told by MGM Studios that the idea was ridiculous as nobody would want to see a giant mouse on the screen. It would scare the women. Over the next fifteen years *The Three Little Pigs* was rejected, *Snow White and the Seven Dwarfs* was initially laughed at and production of *Pinocchio* was halted as nobody could relate to a deceitful, delinquent juvenile boy. *Bambi* was dismissed as inappropriate. Bankrupted several times throughout his career, Walt Disney was even told by the city of Anaheim as they rejected his proposal for a theme park that it would only attract riff-raff. And yet throughout all of his rejection and failures Walt Disney is remembered as the greatest, and most profitable, animator in history, and he died in 1966 a fabulously wealthy man with a store-room full of Academy Awards.

Frederick W. Smith (b. 1944) is the founder and chairman of FedEx, one of the largest courier companies in the world. As a boy he was crippled by a rare bone disease, although he recovered enough to become a keen pilot by the age of fifteen. Whilst attending Yale University, where he studied economics, he wrote an essay outlining his idea for an overnight delivery service in the computer information age. Legend has it that he received a C-grade and was told by his professor that to achieve a higher mark his idea had to be at least feasible. The paper later became the business plan for the world's first, and most successful, overnight delivery company.

Frank Winfield Woolworth (1852–1919) worked as a stockroom boy in a general store as a teenager. He had been told by the owner that he 'lacked the sense' to be allowed to serve customers. Confined to the back room, he had the idea for a shop where every item was priced at only five cents. In 1878 he borrowed $300 to open his first Five-Cent Store, which failed within weeks and closed down. In 1879 he tried again, but expanded his idea to include ten cent items. By 1911 the F. W. Woolworth Company traded from nearly 600 locations, and at the time of his death, in 1919, his company was valued at nearly $1 billion in today's money.

> While theoretically and technically television may be feasible, commercially and financially I consider it an impossibility, a development of which we need waste little time dreaming about.
>
> Lee de Forest, American radio pioneer and inventor of the vacuum tube, in 1926

BUSINESS AND INDUSTRY

CRUDE OIL

During the nineteenth century, as the Industrial Revolution began to reshape Europe and America, there emerged a new and powerful type of businessman, evolving from the Neanderthal and brutish mill and plantation owners of the previous century. The modern barons were smooth and sophisticated, but no less ambitious and single-minded than their predecessors. Real estate, railroads, steel, construction and even fencing became the products of the rich and powerful. But the product of them all, crude oil, was yet to take its place amongst the elite industries by the time the century reached the halfway mark. However, that was to change, and very quickly indeed.

In around 1845 an American businessman and inventor, Samuel Martin Kier (1813–74), noticed that his salt wells were becoming contaminated by an oily substance that was seeping from the rock as his men drilled for the valuable condiment and preservative. To begin with he ordered his workers to dump the offending mess into a nearby canal, but when one of the pools ignited Kier immediately saw a potential use for the 'rock oil', as he named his discovery. To Kier, it was simple. The primary fuel used in all lamps across America until that point had been whale blubber, an expensive and limited

154

resource, and so Kier employed chemists to experiment with his new discovery. By 1848 he had renamed his product Seneca Oil and was selling it as a skin ointment, but it proved to be unpopular and unsuccessful. A second attempt turned out to receive only a slightly better reception, but his 'petroleum jelly' product was soon being shipped across America as it grew in popularity. You have used it too, dear reader, as these days that exact same product is called Vaseline. If you have ever wondered why the label bears the words 'petroleum jelly', well then now you know.

Despite this, Kier remained focused on the idea that the rock oil burned slowly and he saw it as a way to replace the increasingly expensive whale oil. By 1851 Kier had built a refinery and was producing safe, new oil lamps for the mining industry, although with only moderate success. At that time, with limited resources of seeping rock oil, Kier seemed to have become bored with the idea, to the point that he even failed to register any patents for his new products. However, in 1857 two of his executives, George Bissell and Jonathan Eveleth, heard of a rock-oil pool that had formed close to the town of Titusville in Pennsylvania and were sent to investigate. There, in their hotel, they encountered Edwin L. Drake (1819–80), a retired railroad conductor who had settled in the town with his family. Impressed by their new acquaintance, Bissell and Eveleth employed Drake, largely on the strength of his free railroad travel pass, to tour the country investigating other rock-oil pools. By then, they knew there was a market for the product, if only there was enough of it to be found.

In the spring of 1858 Drake travelled to what is now known as Oil Creek, which at the time was only a forty-five mile tributary of the Allegheny River between Crawford and Allegheny counties in Pennsylvania. It was an area that had been drilled before; salt and fresh water had then been the objective. When rock oil appeared nobody really knew what to do with it and so they moved on to new locations. However, having investigated the region and making the sort of geological survey only a retired railroad ticket collector could make, Drake decided to adopt the salt-well drillers' technique of using a steam-engine powered drill for the oil he was convinced lay beneath the surface. This decision led to one of the most famous quotations the oil industry has ever recorded when the foreman of the first team he tried to engage said to him: 'Drill for oil? You mean drill into the ground and try to find oil? You're crazy.' It was exactly the sort of thing drillers had been trying to avoid until then and so the foreman packed up his equipment and left.

A new team was eventually assembled and in the summer of 1859 drilling began, but they soon encountered issues with the loose gravel surface that kept collapsing and stalling the drill bit. Drake overcame the problem by driving a cast-iron pipe into the ground along with the drill, preventing further collapses and water penetration. Even so, progress was painfully slow at a rate of only three feet per day. The drilling team began to lose motivation, crowds of onlookers gathered to jeer at what had, by then, been nicknamed 'Drake's Folly' and, incredibly, the Seneca Oil Company abandoned the project, leaving Drake to his own devices. Funding soon ran out but, with the help and generosity of others, Drake persevered until 27 August 1859 when the drill bit reached a depth of seventy feet and hit a crevice. Once again, the team packed up for the day, but when the chief engineer, Billy Smith, arrived for work the following morning he was amazed to see the rock oil slowly rising up through the shaft. It was quickly hand-pumped into an old bath tub and the date is forever commemorated as the day the 'crazy man first struck oil'.

The following morning other prospectors were using Drake's method of drilling inside a pipe, to prevent collapse (a method still used by drillers to this day). To begin with, Drake was producing forty-five barrels of oil a day, all of which was sent to Samuel Kier's refinery and then sold as lamp oil throughout the United States. Within a decade 16,000 barrels a day were being refined and Drake was being honoured as the father of the modern oil industry. However, his business acumen failed to match his inventive enthusiasm and, having failed to patent his drilling invention and losing all the money he made with poorly judged investments, by 1863 Drake and his family had become impoverished and almost destitute. Finally, in 1872, the Commonwealth of Pennsylvania awarded the old man an annual income of $1,500 in recognition for founding the oil industry.

The business barons who followed – in the shape of Henry T. Ford, J. D. Rockefeller and subsequently most of Texas, and for that matter the Middle East – would never have enjoyed the great power and wealth that Edwin L. Drake initiated when all around him had given up on the idea of finding, or using, rock oil. In fact, without Drake, the world's entire oil and petroleum industry might well have been limited to little pots of Vaseline. And whales may have become extinct too. For his part, Drake died on 9 November 1880 in Bethlehem,

Pennsylvania, and is buried with his wife at Titusville alongside a fine memorial built to honour his memory.

> **How Wrong Can You Be?**
> George Sutherland, American author of *20th Century Inventions*, said in 1900: 'The amount of misguided ingenuity which has been expended on these two problems of submarine and aerial navigation during the nineteenth century will offer one of the most curious and interesting studies to the future historian of our technologic progress.'

CLINTON'S DITCH

In the years following the War of Independence the United States of America began to encourage immigrants from all over the world to travel west and to help populate, and build, the New World. And many more arrived as they had no other choice (see 'The Potato'). As the Napoleonic Wars raged all over Europe, political and religious refugees poured into the ports of the former British colony where they had been promised land, freedom and unlimited opportunity. From the East Coast, settlers pushed west and set up farms and homesteads deeper and deeper into the continent. Within only a few decades settlers had moved so far into the interior that they began to find themselves isolated and out of range. Farmers and fur traders found it could take weeks for their produce to reach the big cities by ox cart. In as early as 1785, the President, George Washington, had been looking for ways to use the Potomac River as a navigable link with the west.

To the north there were several proposals for ways to link the port of New York to the Great Lakes, including one to build a canal from Lake Ontario to the Hudson River. But nobody, it seemed, could justify a need for one, let alone invest in such a project. Jesse Hawley (1782–1842), a flour merchant from Geneva, New York State, had been struggling to move his produce along the dusty cart tracks that made up the commercial trading routes of the day, and in 1807 he found himself bankrupted and in the debtors prison where he would languish for nearly two years. Whilst there, Hawley began writing essays, using the pen name of 'Hercules', extolling the virtues of a man-made canal network that would connect Lake Erie to the Hudson River and subsequently New York City. Eventually, Hawley's articles began to appear in the *Genesee Messenger*, providing detailed and eloquent argument as to why such a project would be of huge value to both the state of New York and the nation at large. Although Hawley's ideas proved credible to some, they were dismissed as the 'ramblings of a madman' by most. And it is easy to see why.

In 1807, with the new nation still under thirty years old and paying the price of independence (in commercial terms), a small-town flour merchant was proposing that the wealthy citizens of New York City pay for an 87 mile (140 km) man-made canal simply to connect a handful of rural communities with the big city. And that was only the beginning. The complete project would cover 360 miles (580 km) and require cutting through solid rock, building viaducts over valleys and existing rivers, and installing more than fifty locks to cope with the 200 metre difference between sea level and Lake Erie. President Thomas Jefferson rejected the plan as 'a little short of madness', although New York Governor DeWitt Clinton (1769–1828), who himself had ambitions of becoming president, had other ideas. Despite much opposition, ridicule and even threats, Clinton managed to persuade the Senate to approve the project and provide a budget of $7 million, an astronomical sum in 1807. The newspapers immediately dismissed the idea and dubbed the project 'Clinton's Big Ditch' or 'Clinton's Folly'. In other words, it was a complete waste of money and a pure indulgence of the Governor alone. The citizens of New York were furious.

Despite such opposition, and the fact that there was not a single qualified civil engineer even living in the United States at the time, work began, on 4 July 1817, on the largest construction project the Western world had embarked upon in over 4,000 years. One young amateur by the name of Canvass White had persuaded Clinton to allow him to travel to Britain and study the canal network. He returned with enough understanding of locks and viaducts to be of great value to the overall project. Against the tide of public opinion, DeWitt Clinton saw the canal as the key to making New York the most important city in America by linking the Atlantic Ocean to both the interior and the Great Lakes. It was a vision that would soon turn the whole region into the economic power-house of America. The Erie Canal was completed in less than eight years by 50,000 labourers using nothing but pick, shovel and minor explosives. Most of them were Irish or Chinese immigrants earning more than five times the money they could have expected back home. But it was hazardous work and over a thousand would lose their lives in the process.

Although the canal was opened for trade as each section was completed, the formal opening of the entire project was held, with great ceremony, on 26 October 1825. It was a miracle of engineering and an instant economic success. $15-million worth of goods were transported along its route during the first year alone, the equivalent of nearly $300 million in today's money. And it all moved along twenty times faster than it had taken by ox-cart and dirt road. Food prices for the cities crashed by 95 per cent. Major towns and new cities sprang up along the way in the shape of Syracuse, Rochester and Buffalo. The frontier people, who had become used to being self-sufficient, were now able to buy anything they wanted, from anywhere in the world; and New York City, as DeWitt had predicted, became a boomtown. Wall Street was established as the financial centre of the Western world and the city immediately became the nation's number one port. So much money flowed through New York, as a result of the canal, that the word 'millionaire' was invented in 1840, only fifteen years after it had first been opened.

Jesse Hawley's fortunes also changed and in 1820 he became a member of New York's State Assembly. DeWitt Clinton never realized his ambition of becoming president, as he died suddenly in 1828, but he did live long enough to see the tide of public opinion swing firmly in his favour. Despite previous criticism and ridicule, as soon as the canal was complete the New York newspapers were celebrating his achievements. This, of course, may have had something to do with their new, and massively increased, circulations. DeWitt Clinton would possibly have finally felt appreciated by his beloved city when the *New Hampshire Sentinel* published an article which read:

161

'The efforts of Governor Clinton, to advance the best interest of the State over which he presided, are very generally acknowledged both by his constituents and the public abroad. His exertions in favour of the great Canal have identified his name with that noble enterprise, and he will be remembered while its benefits are experienced.' The piece ended with the words, 'Yield credit to Clinton, and hail him by name.' A sentiment the great man would surely have appreciated if he had still been alive when it was finally published.

UNZIPPED – THE TRUE STORY OF THE ZIP

As the Industrial Revolution in Europe and America gathered pace during the nineteenth century, in line with mass migration from the Old World towards the land of opportunity, one man set about changing the way the manufacturing of clothes was being handled. On 10 September 1846 Elias Howe (1819–67), an apprentice mechanic, filed and was awarded a US Patent for the first mechanical sewing machine, using a lock-stitch design. His invention included three crucial features that are still found in every modern-day sewing machine: an automatic cotton feed, a shuttle that operated beneath the material that formed the lock stitch and, vitally, a needle with the eye at the tip, instead of at the top as was common with handheld sewing needles. His family would later claim that he came up with the design whilst dreaming. When he woke, the story goes, Howe ran to his workshop, scribbled out a diagram of the idea and the mechanical sewing machine was born. However, Elias was unable to attract funding for his new device and had to travel to England in 1847 where he sold his first machine, for

£250, to William Thomas, a corset and umbrella maker of Cheapside in London.

But that was it. Penniless and unhappy, Howe returned to Cambridge in Massachusetts to find his beloved wife Elizabeth seriously ill. She died shortly after he arrived. To add to his trauma Howe also found that Isaac Singer had managed to create an exact copy of his design and was having considerable success in mass marketing and distributing his Singer sewing machine. Howe launched a legal action and then returned to his work-bench to begin work on some of the other ideas he had up his sleeve, one of which was the Automatic Continuous Clothing Closure for which he received a patent in 1851. This device featured a line of small, inter-locking metal clips that would later become known as the zip. However, Howe was otherwise distracted by his legal disputes with Isaac Singer and made very little effort to develop or market his invention. In 1854 he won his case against Singer and began to earn considerable roy-alties. Despite contributing much of his fortune to the Union Army during the American Civil War (1861–5), Elias Howe died a very rich man in 1868, having never revisited his Automatic Continuous Clothing Closure device.

As the war drew to a close and the national rebuilding programme began, Americans took to wearing the high button leather, or rubber, boot, largely as a result of the ankle-high mud and horse manure most of them had to pick their way through, even on city streets. In 1893 American inventor Whitcomb L. Judson (1846–1909) modified Howe's original design, after realizing it pro-vided an easier way of putting on and taking off his high button boots. A fat man, Judson was fed up with the ordeal of bending over to button and unbutton his boots every day and so he formed the Universal Fastener Company to distribute his new product, which he called the Clasp Locker. Debuting at the Chicago World's Fair in 1893 as a boot fastener, the Clasp Locker met with little success. Boot manufacturers at the time, it seemed, were quite happy with their cheaper lace and button system. Judson, however, was convinced of the benefits of his new device; he reorganized his company into the Fastener Manufacturing and Machine Company, moved to New Jersey and in 1906 hired Gideon Sundback, an electrical engineer of Swedish origin, to work for him. Sundback could also see the value of the Clasp Locker and set about improving the design, and in 1914 a patent for the new Separable Fastener was awarded.

Judson was convinced of its impending success and was quoted as saying, 'From the foregoing statements it must be obvious that a shoe equipped with my device has all the advantages peculiar to a lace shoe, while at the same time it is free from the annoyances hitherto incidental to lace-shoes on account of the lacing and unlacing required every time the shoes were put on or taken off the feet and on account of the lacing strings coming untied. With my device the lacing strings may be adjusted from time to time to take up the slack in the shoes, and the shoes may be fastened or loosened more quickly than any other form of shoe hitherto devised, so far as I am aware.' And yet, incredibly, still nobody was interested in his invention.

However, in 1923 the B. F. Goodrich Company, who would later become the largest rubber-tyre business in the world, decided to use the Clasp Locker on their new design of rubber boot. Their marketing department came up with the new, modern term of the 'Zipper Boot' and in 1925 they registered 'Zipper' as a trademark. For the next decade the zipper was used exclusively for rubber boots and waterproof tobacco pouches, until an advertising campaign for a new line of children's clothing, featuring the zipper that made it easier for young children to dress themselves. By 1937 French fashion designers and magazines were featuring zippers on men's trousers and jackets, and within another twenty years, a full century after Elias Howe's original invention, the zipper had transformed from a novelty item into the world's most commonly used fastener, with over a million miles being produced every year.

Sadly, it seems, not everybody is happy with Elias Howe's innovation. According to a study published by the *British Journal of Urology International*, zippers are the most common cause of serious genital injury and in excess of 2,200 people suffer zipper-related genital damage requiring hospital treatment every year. (Ladies, notice how they are careful to use the word 'people' here and not only men.) For the record, the second biggest threat to your genitals is the bicycle.

How Wrong Can You Be?

A military adviser to Field Marshal Douglas Haig, at a tank demonstration in 1915, told him that 'the idea the cavalry will be replaced by these iron coaches is absurd. It is little short of treasonous . . . Shoot the designer. Shoot him at dawn along with all the other traitors!' (No, he didn't say that last bit. I made it up.)

THE BRA

There are a couple of interesting issues relating to the invention of the modern bra. The first is that the original patent was submitted on 12 February 1914 to the US Patent Office by twenty-three-year-old Mary Phelps Jacobs (1891–1970). Mary's story is an interesting one in itself, to which we will return, but what really caught my attention was the date of the invention. Now, working on the safe assumption that women have always had breasts, the question becomes: what happened before 1914?

Naturally, women have been wearing support since time began. The earliest written records we have, which can be attributed to the ancient Greeks, reveals that women wore a specialized garment that was designed specifically to support their breasts called an *apodesmos*, which translates as 'breast-band'. It is also known that Roman woman wrapped bandages across their breasts when they were engaging in anything strenuous like sport or war. In recent years archaeologists working in Austria discovered four garments that were made up of two linen cups with both shoulder and torso straps, which have been carbon-dated by scientists to somewhere between 1440 and 1485. And we also know that Catherine de' Medici, the wife of King Henry II of France, banned what she described as 'thick waists' from her court during the 1550s – a move that led to over 300 years of noblewomen forcing themselves into tight whalebone corsets that were firmly laced up at the back.

Whilst fashions changed over the years, the suffering went on until the early twentieth century when the nineteen-year-old socialite Mary Phelps Jacobs attended her first debutante ball in 1910. Now, by all accounts young Mary did not enjoy her first outing at all as she found the restrictive whalebone corset both uncomfortable to wear and also unable to accommodate her particularly large breasts. The end result, it has been noted, was that Mary spent the evening with what looked like one giant boob attached to her chest. Mary was determined to avoid such a spectacle for the second time and, with the help of her maid, designed a simple support made from two silk handkerchiefs, secured by pink ribbon, which managed to hold her in shape for the entire evening of her second ball. Within weeks Mary was making similar garments for her friends and relatives and, by November 1914, had filed a patent for her 'backless brassiere'.

Then, in what might be one of the greatest moments of good timing throughout the history of ladies underwear (indeed, industry itself), the US War Industries Board called for women to stop wearing corsets so that the metal used could be put to use in other ways as the First World War broke out in Europe. At this point the Warner Brothers Corset Company in Connecticut stepped in and bought Mary's patent for the princely sum of $15,000, approximately $375,000 in today's money. At only twenty-four years old, Mary Jacobs was a wealthy woman, although Warner Brothers' investment paid off handsomely as they cashed in to the tune of over $15 million over the next two decades. (Incidentally, the corset boycott contributed over 28,000 tons of steel to the war effort. Enough metal to build two battleships.)

The final part of the tale involves Mary herself as she divorced her alcoholic husband soon after he returned from the war; she then began an affair with a young man called Harry Crosby who was seven years her junior. The resulting scandal led to the couple moving to Paris where they enjoyed the high life on Mary's income and Harry's $12,000 a year trust fund. There they lived in an open marriage, with both having numerous affairs, and together formed the Black Sun Press, the publishing company that introduced young unpublished writers in the shape of James Joyce, Ernest Hemingway, D. H. Lawrence, T. S. Eliot and Ezra Pound, to name but a few. For her own part, Mary lived a long and decadent life before finally succumbing to pneumonia in 1970, having witnessed the famous bra-burning campaign that was associated with the women's rights movement of the 1960s. No doubt Mary would have approved.

> Fooling around with alternating current is just a waste of time. Nobody will use it, ever.
> American inventor and businessman
> Thomas Edison, attempting to mock his rival,
> George Westinghouse, in 1889

IT WAS A CLOSE SHAVE

King C. Gillette (1855–1932) was a creative entrepreneur and from the age of seventeen he dreamed of becoming an inventor and making his fortune. But there is quite a long way between dreaming of something and then actually inventing a brand new product. He was born in 1855 and Gillette's parents, both successful inventors themselves, then moved the family to Chicago where their business was devastated in 1871 during the Great Chicago Fire. For the next thirteen years Gillette struggled to make a living from his inventions and, in the end, managed to fail miserably.

Eventually he found work as a salesman, although he enjoyed little success and finally returned to his parents' house at the age of forty, unable to pay his bills and with few future prospects. But Gillette refused to give up and went to work for William Painter, a local entrepreneur who had invented a disposable bottle cap, which he had turned into a successful business. In 1892 Painter had been awarded a US Patent for his product, formed the Crown Cork and Seal Company and employed King Gillette to walk the streets of Baltimore and sell the bottle tops from door to door.

One day, whilst Gillette was out on his rounds, Painter joined him and explained that the secret to a successful product was to invent something people would use once and then throw away. For Painter, the cork-sealed bottle top had proved to be the perfect product. Gillette then decided he needed one of his own. The following weeks turned into months and Gillette spent every waking moment trying to think of a product currently used by everybody that he could make disposable. One morning Gillette was shaving at his basin when he cut his face with his old, blunt cut-throat razor and, as his blood dripped into the bowl, his search for an invention was over. He realized there and then that instead of every man sharpening an old straight-bladed razor on a leather strap every morning, he would design thin, sharp blades that were so cheap to produce they could be thrown away after only a few shaves. That morning he left a note for his wife, reading, 'I've got it, our fortune is made.'

Gillette then spent all of his spare time trying to find a way to manufacture blades out of thin steel that was strong enough not to bend or buckle, without success. Eventually Gillette, tired of his friends and associates

169

joking with him about his failure in creating a new and modern safety razor, turned to Steven Porter, a machinist from Boston. Whilst he was eating a sandwich. Porter thought up a way of sandwiching a thin, sharp blade between two stronger pieces of steel, leaving only the sharp, leading edge exposed. In the summer of 1899 Gillette became the first man to shave with a disposable bladed razor. But public opinion was firmly against him as, historically, a man's razor was considered to be a one-off purchase for life. Often razors were handed down from grandfather to grandson, and the idea of throwing one away after only a few shaves seemed alien to the general public. Financing a marketing and advertising campaign was also out of the question as Gillette himself was broke and potential investors were sceptical.

However, William Emery Nickerson, an expert machine inventor, took a look at the razor and realized that if he made the blade removable, by adding a screw-fix device, then only the thin, cheap blade needed to be thrown away and not the whole razor. Gillette immediately filed for a patent and formed the American Safety Razor Company in partnership with Nickerson and two investors. The name of the company was soon changed to the Gillette Safety Razor Company, and when early prototypes impressed investors Gillette was offered $125,000 in exchange for 51 per cent of the company by a group of New York investors. The first advertisement for the Gillette safety razor appeared in *System* magazine in early 1903 offering one razor and twenty blades for $5, which was around half the average man's weekly wage at the time. Needless to say, sales were slow and by the end of that year only fifty-one razors had been sold by mail order. Gillette remained a salesman for the Crown Cork and Seal Company and was quoted later as saying, 'The razor was looked upon as a joke by all of my friends. If I had been technically trained then I would have quit there and then.'

But the other investors remained positive and, despite nobody having drawn a salary from the new company, they decided to reduce the number of blades offered with the razor from twenty down to twelve. With that, Gillette resigned and travelled to England, but when he heard that his partners were planning to sell his patent to a European company he raced back to Boston and convinced his allies to allow him to regain control of the company. With new investment and renewed enthusiasm Gillette embarked on an advertising campaign that led to sales, in 1904, of 90,884 razors and 123,648 blades. By 1908 the company had established factories in Germany, France, Britain and Canada, and had sales exceeding 450,000 razors and 70 million blades. As the First World War broke out Gillette read that French and British troops needed to remain clean shaven in the trenches so that their gas masks would seal properly. When America entered the war Gillette offered the American government razor sets for every soldier at cost price and received an order for 3.5 million Gillette safety razor kits.

This firmly established the reputation of the company and ensured a generation of men's loyalty to its products. By then Gillette had become a multi-millionaire and all but retired from the company bearing his famous name. But, unfortunately, he invested most of his money in property and on Wall Street, and lost the bulk of it during the infamous stock-market crash of 1929 and subsequent Great Depression. Sadly, the man whose company would

171

be sold in 2005 for $57 billion died alone, a virtual bankrupt.

> That is the biggest fool thing we have ever done [researched]. The bomb will never go off, and I speak as an expert in explosives.
>
> William D. Leahy, US Admiral, advising
> President Truman on atomic weaponry in 1944,
> a year before Hiroshima and Nagasaki

KITTY LITTER

At the end of the Second World War, Edward Lowe (1920–95) was discharged from the US Navy and then found his family were so badly off that he had to hitch-hike his way home to Cassopolis, Michigan. There his father ran a business supplying sand and sawdust to local mechanics and farmers to soak up spills on their workshop floors. It was a modest company and many of his customers found it a luxury they could not afford during the war years. Upon his son's return, Henry Lowe handed over the delivery business and went off to concentrate on the family's tavern in Vandalia, a few miles along the road.

With the responsibility of a wife and young child Ed Lowe set about expanding the family firm by building the customer base and introducing new products, one of which was a super-absorbent clay called fuller's earth. Lowe, however, was frustrated to find that his clients were just not interested in the more expensive clay product and he was left with tons of the stuff that he could not shift for free. However, in January 1947, Ed was returning home just as his neighbour, Kaye Draper, was trying to dig sand out of the frozen ground to line her cat box in the

172

basement. In those days it was common for families to put their cats out for the night as, being desert animals originally, their water retention led to highly concentrated, undiluted and smelly waste. During the winter months many would put an old vegetable tray filled with sand, sawdust or earth in their basements so their pets could at least stay out of the freezing cold weather. Kaye approached Ed and begged him for some sawdust but, instead, Lowe shovelled a couple of scoops of fuller's earth, from the heap in his garage, into her cat tray. He didn't know what else he was going to do with the stuff and was trying to get rid of it.

The following morning Kaye was amazed to find that not only did her cat feel comfortable in the clay, but that its absorbent qualities meant there were no lingering smells. She also noticed that the fine grain clay didn't stick to her cat's feet and leave a trail throughout the house, as sand and sawdust always did. The following day Kaye approached Ed with a handful of her cat-loving friends to explain how good his clay was and Lowe immediately saw an opportunity. That afternoon he bagged five-pound sacks with fuller's earth, wrote the words 'Kitty Litter' in black ink on the front, piled a dozen of them into the back of his truck and set off for the nearest pet stores.

However, Ed found that he met with the same resistance from the store owners as he had from his factory and workshop clients. Nobody, they told him, would pay sixty-nine cents for a bag of fuller's earth when sand and sawdust was half that price and soil was free. In the end, frustrated and downhearted, Ed left a couple of bags with each store and suggested they gave it away for free. Within a week store owners were surprised to find cat

173

lovers returning to ask for Kitty Litter, and even more surprised that they were prepared to pay for it. Ed soon started receiving orders and, encouraged by this response, began loading his truck and travelling all over the county repeating his sales model of leaving a couple of bags for free and then waiting for the orders to come in.

Meanwhile, he constantly improved his product and packaging, offering money-back guarantees, introducing deodorizers and scents, and including handy plastic scoops. He was soon travelling the country demonstrating his product at cat shows and county fairs throughout the land. By then he had realized that everyone who used his Kitty Litter product always came back for more, so he took a calculated risk and bought a fuller's earth processing plant before exporting his Kitty Litter around the world.

How Wrong Can You Be?
Soviet Premier Nikita Khrushchev confidently predicted the outcome of the global conflict between communism and capitalism when he warned America and the West, in 1958, that 'we will bury you'.

His company began to employ scientists at a modern research and development centre who worked continuously to upgrade existing products and develop new ones. Attached to it was a cattery where Ed gave homes to over 120 strays that helped the scientists with their progress. He also built a fully staffed care clinic and installed 24-hour security. Ed Lowe hadn't just invented a product, he had invented a worldwide industry, and when he retired in 1990 his company was enjoying annual retail sales of $210 million. Today Kitty Litter is thought to be worth half a billion dollars, and the product everybody told Ed Lowe that nobody wanted is part of an £11 billion a year industry, in America alone.

BALLPOINT PEN? WHAT'S THAT, THEN?

Prior to the 1880s, the only way to write anything down onto paper was to use something that could be fashioned into a nib, such as a feather or piece of wood or shell, and dip it into an inkwell. Or use a pencil. That is, unless you had one of those fountain pens that had become popular with the rich and famous, but many of them still had to be dipped in ink until the 1880s when fountain pens began to be mass produced for the first time. However, if you were a humble tradesman you probably couldn't afford one of those. Unless you were a tradesman such as the leather tanner John J. Loud (1844–1916), for example.

During the early 1880s John J. Loud, from Weymouth, Massachusetts, had been experimenting with ideas for things he could use to mark his leather products, without much success. His fountain pen often failed to work and so Loud set about designing a pen that had a thin metal tube with a tiny, rotating steel ball held in place by a small burr. His idea was to fill the tube with ink, let it coat the ball and continue to do so as it rolled across the leather, leaving ink marks in whatever pattern he chose to make, including words, obviously. Loud was excited by his design and even filed a patent for a 'roller ball tip marking pen', which was awarded on 30 October 1888 in his own name.

175

The problem Loud faced was that whilst his design worked adequately on leather, and other rough surfaces, it was not so efficient on smooth surfaces, such as paper. Loud realized that with a little development he could refine his design but was told by everybody he approached that he was wasting his time. After all, the perfect pen had already been designed and was now in mass production. Why, fountain pens even delivered their own ink in those modern times and there was no more need for dipping, or inkwells, or any smudges. 'You are too late, Mr Loud,' they told him, 'we already have the perfect pen thank you.' And with that, John L. Loud went back to stitching handbags, or shoes, and no more was heard of him.

No more was heard of his idea either until 1935 when a Hungarian newspaper editor was becoming annoyed by the amount of time he wasted filling up his, by then, old-fashioned fountain pen. He was also fed up with clearing up ink smudges and his nib tearing through the news-print. But he noticed how the ink from his newspaper press dried ten times faster than that of his pen and so Ladislas Biro (László Bíró, 1899–1985), with some help from his brother Georg, a chemist, set about finding a solution. Over the following few years Ladislas experimented with ball-tipped designs of the exact nature John J. Loud had patented all those years earlier, whilst Georg developed ink samples using the thinner, lighter inks of the printing press. During the summer, whilst taking a break from their work, they were at the seaside where they met a fascinating old gentleman who loved their model of a ballpoint pen. The old man turned out to be Agustín Pedro Justo, the serving President of Argentina, who urged the brothers to move to his country where he would help them fund a factory.

The following year, as war broke out in Europe, the brothers did just that and fled to Argentina, stopping off in Paris to file a patent along the way. Once they had settled in Argentina, they found no shortage of investors and established a factory in 1943. But they found that their new pen didn't work very well at all, and they had to go back to the drawing board and refine their design. The second version faired a little better but sales throughout the country did not meet their expectations and eventually the money ran out – although not before American pilots, who had been stationed in Argentina during the war, returned to America full of enthusiasm for the new pens that worked perfectly at altitude and did not need refilling very often.

The US Air Force sent specifications to a handful of American companies and one of them, in an attempt to corner the market, paid the Biro brothers half a million dollars for the US manufacturing rights to their patent. Meanwhile, a Chicago salesman called Milton Reynolds, who had bought several Biros whilst on holiday in Argentina, thought he could avoid any legal problems because the original patents had expired and so he set about copying Biro's design (or Loud's, depending upon your viewpoint) with sufficient improvements to allow him to obtain his own US Patent.

Reynolds then showed his prototype to his friend Fred Gimbel, whose family owned the Gimbel's department stores, which were, at one time, the largest store chain in the world. Gimbel arranged a clever marketing campaign and launched the new ballpoint pen in New York City on 29 October 1945, just two months after the end of the

Second World War. Priced at $12.50, the cost of a single night in a decent New York City hotel room, Gimbel's sold out their entire stock of 10,000 within two hours as 5,000 people crowded into the shop. The New York Police Department had to despatch fifty officers simply to maintain crowd control. Over the following six weeks the Reynolds International Pen Company worked around the clock to make eight million pens in order to keep up with demand. Reynolds became a very wealthy man and even bought a magnificent French estate, the Château du Mesnil St Denis, to use as the headquarters of his European division. But Reynolds was an astute businessman and realized that other companies would soon be flooding the market with cheaper versions of their own ballpoint pens, and so he sold his company and retired to South America in 1947.

> There is not the slightest indication that nuclear energy will ever be obtainable. It would mean the atom would have to be shattered at will.
> Albert Einstein, German-born American physicist,
> in 1932

THE RAILWAY NETWORKS

The first to adopt the idea of moving heavy loads on railed tracks were the ancient Greeks, followed by the Romans, who both used stone tracks to move loads in wagons, pulled by animals, from their quarries. The Greeks even had a paved track across the Isthmus of Corinth, along which they hauled their ships between the Ionian Sea to the Aegean Sea to speed up their naval campaigns during the sixth century BC. The paved track was in use for 500 years. It took another 1,500 years before narrow-gauge wooden railways became common in mines throughout Europe; during the eighteenth century, iron railways were laid for the first time for public use and the world's first horse-drawn public railways began to appear.

With the development of the steam engine during the Industrial Revolution in England, audacious plans were soon being made for a locomotive-hauled train that could carry passengers, freight and mail along permanent tracks all over the country. By 1829 George Stephenson (1781–1848) had successfully demonstrated his *Rocket*, largely designed by his son Robert, and the following year, on 15 September, the world's first intercity railway was opened between Liverpool and Manchester. This was also the day of the first railway fatality when the MP for Liverpool, William Huskisson, disembarked during the opening ceremony to approach the Duke of Wellington. Huskisson failed to see the *Rocket* speeding towards him on the adjacent track and died from his injuries later that evening.

Meanwhile, within the scientific community, the brilliant minds of their time were strongly warning against the development of a railway network and, in particular, the use of speed. Dionysius Lardner (1793–1859), an Irish scientific writer and early critic of the steam train, warned in 1828, 'Rail travel at high speed is not possible because passengers, unable to breath, will die of asphyxia.' And he wasn't the only one. French doctors predicted the spread of disease as 'the rapid transition from one climate to another will have a deadly effect and a sudden change in diet can cause dyspepsia and dysentery'. Dr John Keate (1773–1852), the distinguished headmaster of Eton College, personally wrote to his former pupil, the newly elected Member of Parliament William Gladstone, urging him to use his personal influence to prevent the construction of the railway line into Windsor. Keate complained that it would 'interfere with the discipline of the school, the studies and amusements of the boys, affect the healthiness of the place, increase of floods, and endanger even the lives of boys'. Windsor and Eton Station was eventually opened in 1849.

And it didn't end there. A group of notable scientists warned about the 'spectre of pleurisy that would certainly attack passengers when going through tunnels, always provided that they escaped the catastrophes resulting from the explosion of locomotive boilers'. And a British doctor issued a warning that train travel would expose otherwise healthy people to colds and consumption. In London the influential *Quarterly Review* asked, 'What can be more absurd than the prospect held out of locomotives travelling at twice the speed of stagecoaches?'

In America, the Governor of New York, Martin Van Buren, wrote to the President Andrew Jackson in 1830, while he was Secretary of State, claiming, 'The canal system of this country is being threatened by a new form of transportation known as railroads. As you may know, Mr President, railroad carriages are pulled at the enormous speed of 15 miles per hour by "engines" which, in addition to endangering the life and limb of passengers, roar and snort their way through the countryside, setting fire to crops, scaring the livestock and frightening women and children. The Almighty certainly never intended that people should travel at such breakneck speed.' The enlightened Van Buren would succeed Jackson to become the eighth President in 1837.

Dionysius Lardner and the rest were soon proved to be wrong and within twenty years there was over 7,000 miles of railway track networking across Great Britain. But it would not be his final clash with the growing railway network. Whilst Isambard Kingdom Brunel (1806–59) was building the Great Western Railway during the 1830s, Lardner criticized Brunel's famous box tunnel between Chippenham and Bath, and insisted that if the train's brakes were to fail then the gradient from east to west would accelerate the carriages to over 120 miles per hour. At that speed, he claimed, passengers would suffocate. Brunel pointed out that Lardner's calculations failed to consider friction and air resistance and so he had to be wrong.

In 1836, when Brunel proposed to build the first transatlantic steam ship, the SS *Great Western*, Lardner addressed a meeting of the British Association for the Advancement of Science in London and insisted that attempting a sea voyage directly to New York from Liverpool would be impossible and they may as well propose a journey from Liverpool to the moon. The Irishman claimed the ship would run out of coal after 2,080 miles (3,350 km). As it turned out, the SS *Great Western* steamed into New York with 200 tons remaining. Dionysius Lardner himself was no stranger to scandal and controversy. In 1840 news of his affair with Mary Spicer Heaviside leaked out and the cheating pair fled to Paris. The lady's husband, Captain Richard Heaviside, tracked the couple down and horse-whipped Lardner in public. Back in London he successfully sued the scientist for criminal conversion (adultery) and received compensation of £8,000. The Heavisides divorced in 1841 and the furtive pair were themselves married five years later.

181

The scandal effectively ended Lardner's career in London and so he remained in Paris until shortly before his death in 1859. No doubt to the relief of Isambard Kingdom Brunel, and everybody else.

How Wrong Can You Be?
Dominique François Arago (1786–1853) claimed, 'Transport by railroad car would result in the emasculation of our troops and would deprive them of the great marches, which have played such an important role in the triumph of our armies.'

THE HORSELESS CARRIAGE
The first mass-produced motor car rolled off Henry Ford's innovative production line in 1908 and, for the first time, the general public were able to make use of an affordable horseless carriage. A century later the number of cars in use around the world is estimated to be over one billion, with 25 per cent of those in America alone, making the motor car the most popular, or common, way of travelling on the planet.

The first self-propelled vehicle in history was designed by Nicolas-Joseph Cugnot (1725–1804), a French military engineer, in 1769, and his steam-driven machine was used at the Paris Arsenal to move heavy cannon around. However, with a top speed of only 3.7 mph (6 km/h) it would be a long while before the horse had anything to worry about. Cugnot's second effort was a little faster and made history by becoming the first motorized vehicle to ever crash after it was driven into a wall. In 1807 Swiss designer François Isaac de Rivaz (1752–1828) filed a patent for the first hydrogen-powered combustion engine, which he eventually fitted to a six-metre-long chassis weighing over a ton.

It would be another fifty years of tinkering and patent filing before the Belgian Jean Joseph Étienne Lenoir (1822–1900) produced a viable self-propelled vehicle, and by the end of 1870 there were 500 of them running around in Paris. Although, with a maximum speed of only 19 mph (30 km/h), the horse was still a long way from being replaced as the favoured method of transport. There had been progress in Britain, too, as indicated by the Red Flag Act of 1865, which required all vehicles to be operated by three people: one to steer, one to stoke the boiler and one to walk fifty yards up the road with a red flag to warn other road users that it was coming and to let the driver know when he had to stop.

By the time Ford's Model T rolled off the production line in October 1908 there were already more than 10,000 cars creating dust clouds across America, but not everybody was as enthusiastic as the legendary old car maker. In 1909 *Scientific American* magazine reported that 'the automobile has practically reached the limit of its development is suggested by the fact that during the last year no improvements of a radical nature have been introduced'. Ford noted that no improvements were necessary, and by the time the Model T went out of production in 1927, over fifteen million of them had had been sold.

When Henry Ford incorporated the Ford Motor Company in 1903 he appointed his local solicitor, Horace Rackham of Rackham and Anderson, to complete the paperwork. He also encouraged the partner to buy shares in the company and Rackham turned to his friend, the president of the Michigan Savings Bank in his home town, for advice, who told him, 'The horse is here to stay, but the automobile is only a novelty, just a fad.' Against this advice Rackham borrowed money, sold some parcels of land and raised $5,000, which he used to buy fifty shares and become one of only ten shareholders in the new company. Two others were Horace and John Dodge, who later went on to form their own car company in 1915. Within five years the partners were earning more from Ford stock dividends than they could make as lawyers and so Rackham closed his practice and became the full-time chairman of the Ford Motor Company. In 1919 Henry Ford acquired Rackham's entire stock-holding for $12,500,000, at which point he promptly retired and spent the rest of his life giving money away to children's charities. At the time of his death in 1933, Horace Rackham was still worth around $17 million, thanks to ignoring the advice of his bank manager.

Not all of the gloomy predictions about the impending failure of the car industry turned out to be inaccurate. In 1899 the *Literary Digest* had loftily announced that the 'horseless carriage is at present a luxury for the wealthy and although its price will probably fall in the future it will never, of course, come into as common use as the bicycle'. In fact, until 1965 the world production of cars and bicycles remained around the same, although by 2004 over 150 million bicycles were being sold each year, nearly three times the number of cars.

If excessive smoking actually plays a role in the production of lung cancer, it seems to be a minor one.
W. C. Hueper, National Cancer Institute, in 1954

VELCRO

In 1941 Swiss agricultural engineer George de Mestral (1907–90) returned from walking his dog in the Alps and was irritated to notice hundreds of burdock seeds clinging to both his trousers and his dog. After removing them curiosity led de Mestral to place a few under a microscope, and he noticed that the seeds had evolved to include hundreds of minute hooks that would catch on anything with a loop, such as animal fur, to help the plant spread around the countryside. Ever the engineer, de Mestral realized this marvel of nature may also be a way of temporarily, and securely, binding two materials together, and he set about finding a way to recreate the loops and hooks on two different items.

For ten years de Mestral worked away in his spare time but could find nobody interested enough in his idea to offer any support. Eventually he travelled to Lyon in France, the centre of the weaving industry, and managed to persuade one company to manufacture two cotton strips, one with hooks and the other with loops, that actually worked, but the cotton turned out to be far too weak and they tore apart in no time at all. Once again de Mestral was without ally but his prototype had worked, encouraging him to apply for a patent in 1951, which was eventually granted in 1955. With renewed enthusiasm the engineer turned to the much stronger nylon, but he failed to find a way of lining up the hooks with the loops and so his design didn't work. And then, one day, whilst he was on the verge of giving up, de Mestral gave his invention one final throw of the dice. Taking two strips of nylon with the loops, he thought he could simply trim one side with a pair of shears, to make hooks, and see if they then lined up.

It worked and de Mestral had found a way to manufacture his product. Within a year he was ready to go to market. However, the enthusiasm he was expecting failed to materialize as his Velcro product (named after the French words '*velours*', meaning velvet, and '*crochet*', meaning hook) simply looked like leftover rolls of material. And besides, the zip fastener (see 'Unzipped – The True Story of the Zipper') and shoe lace had already been invented and they both worked perfectly well. Undeterred, de Mestral took his product to America where he met with a similar reaction. However, as he continued to look for uses, he found the growing aerospace industry could design their space suits using his Velcro fastener to help astronauts in and out of them. The diving-suit manufacturers soon followed and before long Velcro was being used on ski suits and other sports equipment.

Initially marketed as the 'zipperless zip', Velcro eventually began to be used in children's clothing, and by the mid-1960s de Mestral's factory was producing over 37,000 miles (60,000 km) a year. The inventor enjoyed a near monopoly on his product until 1978, when he forgot to renew his patents and a mass of cheap imitations immediately flooded the market from China and South Korea. By then, however, its original inventor was honoured in his own country and was later inducted, posthumously, into the America National Hall of Fame in 1999. And his Velcro invention had earned him in excess of $100 million. His invention is one of the most versatile the world has ever seen and can be used in thousands of ways between heart surgery and dog coats. George de Mestral's dog, we salute you.

> Transmission of documents via telephone wires is possible in principle, but the apparatus required is so expensive that it will never become a practical proposition.
>
> Dennis Gabor, British physicist and author of *Inventing the Future*, in 1962

E-COMMERCE – THE WORLD AT YOUR FINGERTIPS

In 1974, ten years before home computing made the transition from science fiction to reality, Arthur C. Clarke (1917–2008) was interviewed by Australia's ABC Network and asked how computers, which at that time were large enough to be housed in rooms around the size of a basketball court, would change the future for the everyday person. The author replied with an accurate description of what we today recognize as online shopping and banking. Responding to a question about how life might be for the interviewer's son, Clarke responded, 'He will have, in his own house, a computer terminal through which he will be able to talk and access any information he needs for his everyday life. Like his bank statements, his theatre reservations and anything else required in the course of living in a complex modern society. And this will be in compact form in his own home, which he will take for granted as much as we do the telephone.'

For many, in 1974, the idea that everybody would be able to carry around, in their pocket, information about anything that has ever happened, throughout history, and be able to access anything of current value, would be far too hard to imagine. Remember, in 1974 the *Encyclopaedia Britannica* was made up of over twenty volumes that would be hard enough to transport around in a small van. And even they only scratched the surface in terms of information. It would take people of vision to change our established way of doing things, and yet sometimes even the most imaginative can be embarrassed by their own limitations. For example, in 1966 the esteemed *Time* magazine insisted that remote shopping would flop because 'women like to be able to get out of the house, to be able to handle the merchandise and be able to change their minds'.

In 1986 American technical author Clifford Stoll (b. 1950) was working as a systems administrator at the Lawrence Berkeley National Laboratory in California when he identified an intrusion by the hacker Markus Hess. At that time computer networks were in their infancy and very little was known about intrusion. It is now known that even military networks gave so little attention to their security in those early days that often the default passwords remained unchanged and users were able to log on to many networks by simply typing in the word 'guest'. Stoll, however, was one day given the relatively minor task of finding a seventy-five cent discrepancy on the Lawrence Berkeley network, and he quickly realized that an unauthorized user had logged into the system for nine seconds and not paid for it. Unsurprisingly, recovering less than a dollar was not the motive Stoll had for tracking down the culprit. Instead, he was keen to discover how somebody unknown had gained access to a private network, and why. A ten-month investigation led to a honey trap and, for the first time in history, international law enforcement agencies worked together in order to trace the villain and make an arrest.

Markus Hess, who later confessed to working for the Soviet KGB, became the first person to be jailed as a result of a digital forensic investigation. Clifford Stoll wrote a book about his hunt for the hacker called *The Cuckoo's Egg: Tracking a Spy through the Maze of Computer Espionage* and later contributed to many publications on the subject of online security. Stoll, however, has not always been as successful in identifying internet trends and patterns. On 27 February 1995 he wrote on the subject of growing online communities and noted, 'Visionaries see a future of telecommuting workers, interactive libraries and multimedia classrooms. They speak of electronic town meetings and virtual communities. Commerce and business will shift from offices and malls to networks and modems. And the freedom of digital networks will make government more democratic? That's baloney.'

And he wasn't finished. 'Try reading a book on disc. At best it's an unpleasant chore: the myopic glow of a clunky computer replaces the friendly pages of a book. And you can't take that laptop to the beach. And yet Nicholas Negroponte, director of the MIT Media Lab, predicts that we'll soon buy books and newspapers straight over the internet. Uh . . . sure!'

And all of this was in the same year as internet giants Amazon and eBay were formed. To be fair to Clifford Stoll, this was still four years before any reliable and trustworthy way of sending money over the internet had been developed, and the general public were still a long way from feeling confident about feeding their credit card information into the unknown. Clifford Stoll rounded up his article with the observation that, 'While the internet beckons brightly, seductively flashing an icon of knowledge as power, this non-place lures us to surrender our time on earth. A poor substitute it is, this virtual reality

189

where frustration is legion and where, in the holy names of education and progress, important aspects of human interactions are relentlessly devalued.'

How Wrong Can You Be?

Vannevar Bush, Director of the Office of Scientific Research and Development during the Second World War, once wrote: 'There has been a great deal said about a 3,000 mile high angle rocket. In my opinion such a thing is impossible for many years . . . I say, technically, I don't think anyone in the world knows how to do such a thing. I feel confident that it will not be done for a very long period of time to come. I think we can leave that out of our thinking. I wish the American people would also leave it out of their thinking.'

Well, he was right about that part at least but the rest of Stoll's summary of e-commerce has proved to be inaccurate to the tune of over $1 trillion every year. And this is only the beginning. However, in fairness to Clifford Stoll, it must be conceded that very few people could have foreseen the impact e-commerce would have upon the buying habits of just about every person in the Western world, when considering it at a time when very few people had even heard of the internet, let alone connected to it after unplugging their telephones and using the lead for dial-up access. (It all seems so long ago now, doesn't it?) Stoll, to his credit, can laugh at his previous observations and once noted, when reminded of his *Newsweek* article, 'Of my many mistakes, flubs and howlers, few have been as public as my 1995 howler. Now, whenever I think I know what's happening, I temper my thoughts. Might be wrong, Cliff . . .'

THE POST-IT NOTE – THE ACCIDENTAL BILLION-DOLLAR INVENTION

Spencer F. Silver (b. 1941) studied chemistry at Arizona State University and then earned a doctorate in Organic Chemistry from the University of Colorado before becoming a senior chemist at the industrial giant 3M's Central Research Laboratory in 1966. He was first assigned to a team of five people researching pressure-sensitive adhesive, and their objective was to develop a new, super-strong industrial adhesive. But, instead, Dr Silver accidentally added, by his own admission, more than the recommended amount of the chemical reactant used to cause the molecules to polymerize. The result was not exactly what the good doctor had been trying to achieve but he did notice something unique about his experiment. Unintentionally, Silver had created an adhesive that had what he later described as 'high-tack' but 'low-peel' properties.

Spencer Silver immediately realized he had a developed an entirely new type of adhesive but its possible use was not immediately apparent to anybody. It certainly wasn't what anybody was hoping for as, until then, an adhesive that was weak enough to pull apart was exactly the opposite of what his team was trying to achieve. Undeterred, Silver continued to experiment, although all he was able to produce was an adhesive strong enough to hold two pieces of paper together and yet weak enough to be peeled apart. But it could be reused, multiple times, and that was what appealed to Silver. Years went by and he still could not think of a practical use for his new adhesive. The 3M product developers rejected the idea as they too could not conceive a potential use for it and so, although Dr Silver was frustrated in his efforts, he began to deliver a series of lectures throughout the company in the hope that one of the bright young things might identify a commercial use for his new 'high-tack, low-peel' product. In the process he became known throughout the corridors at 3M as 'Mr Persistent', and for two long years the best suggestion anybody had was to use it as a spray to display company messages on noticeboards without the need for pins.

It was a noble thought but not exactly an idea worthy of international distribution, and the powers that be at 3M remained curious, but unconvinced. Then one day Art Fry (b. 1931), who also worked for 3M in their New Product Development Department, was on the second hole of his local golf course and was told, by a colleague, about Silver's 'interesting adhesive'. Fry worked in the Tape Division Laboratory and decided to attend one of Silver's seminars in order to satisfy his professional curiosity, although he too could think of no practical use for a sticky tape that could be easily pulled apart. At the time there was no demand for that kind of product, and so Art Fry filed the idea away in the back of his mind, where it remained until he had one of those moments of inspiration that most inventors and developers can only dream about.

Five years later, on a Sunday morning and during a particularly boring church sermon, Art Fry, a member of the choir, lost interest and began, instead, to wonder what he could do to stop his bookmarks falling out of his hymn book as he turned the pages. As he pondered the problem he began to recall details of Silver's seminar and started to consider the notion of adhesive, reusable bookmarks. Fry was excited about the idea, and the following morning he made his way straight to Spencer Silver's lab and asked for some samples of his unused adhesive. After several experiments Art Fry tested his new bookmark at choir practice but, although it worked, the residue caused slight damage to the pages. However, after a few more attempts Fry made a bookmark that stuck to the pages and left no mark after it had been removed. Also, each bookmark could be used multiple times. It was the innovative idea Art Fry had been looking for, and so he wrote up his conclusions and presented them to the 3M Development Board. Initially they liked the idea but market research returned poor sales projections and so the adhesive bookmark remained stuck on the shelf, in the back room.

Another couple of years went by until one day, when Art Fry was preparing a report for a supervisor, he wrote a question on one of his bookmarks and stuck it to the front of the report. His colleague wrote the answer on the same piece of paper and returned it stuck to another document and Art Fry had what he would later describe as a 'eureka type head-flapping moment'. He had discovered his use for the easy-peel adhesive: sticky notepads. Fry raced to a neighbouring department where the only paper they could give him was canary yellow and so, for no other reason, the yellow Post-it notepad was born. The samples Fry made and distributed throughout the company proved to be so popular that he later recalled 'executives walking through knee deep snow to ask for replacement pads'.

In 1977, 3M tested the product in four cities under the trade name 'Press n Peel' but the initial sales results were not encouraging. The product, however, remained a firm favourite among 3M staff and so the following year developers decided they needed people to see for themselves how useful reusable, adhesive notepads were and distributed free samples throughout the town of Boise in Idaho: 95 per cent of users soon confirmed they intended to buy the product. That was a good enough return and so, in 1980, 3M finally launched its new, innovative Post-it Notes. They then sold fifty million packs during their first year of trading.

> I have travelled the length and breadth of this country and talked with the best people, and I can assure you that data processing is a fad that won't last out the year.
>> The editor in charge of business books
>> for Prentice Hall, 1957

Within two years Post-it Notes had been established as a necessity and 3M had dedicated an entire production line for that one product alone. They were soon indispensable in schools, libraries, homes, workshops, offices, and were available in all shapes, sizes, fragrances and colours, although the original canary yellow is still, by far, the most popular. In modern times 3M still sells around $3.5-billion worth of Post-it Notes every year, although after their patent ran out, during the 1990s, they are now in competition with many other manufacturers. Dr Spencer F. Silver and Art Fry, for their part, became 'heroes of invention' and have both won the highest honours for research that 3M bestow. They are also the recipients of numerous international engineering and invention awards. In later years Spencer Silver commented: 'If I had thought about it, I wouldn't have even done that experiment. The literature is full of examples that say you can't do this.'

VULCANIZED RUBBER: CHARLES GOODYEAR

Rubber, as a natural resource, was first noticed by Christopher Columbus and his fellow conquistadores as they discovered and began to document the Americas during the late 1400s. It was recorded that the natives played games with elastic balls which appeared to be made from a milky, white sap that was found in local trees and came to be called *Castilla elastica*. They were unknown in Europe at the time. But, to the Europeans, this was only a passing curiosity as the focus of their attention, at the time, was to fill their ships with the gold, silver and jewels that were in such demand on their own continent. However, it was noted that the Olmec tribe (which translates as 'rubber people') skilfully used the dried and hardened resin as a waterproof material out of which they fashioned durable footwear, clothing and water flasks. It was known locally as '*caoutchouc*', which translates as 'the wood that weeps'.

It would be another 200 years before European land surveyors took a closer interest in the natural resource. They realized that the material was difficult to work with and the raw, liquid latex was almost impossible to safely transport. However, during the mid-eighteenth century new scientific developments led to limited experiments and some successful uses of rubber, but it was still an unreliable material that was sticky and smelly in warm weather or hard and brittle during the winter. In 1823 Scottish scientist Charles Macintosh (1766–1843) developed a partial solution when he enclosed a tacky layer of rubber between two strips of material and invented waterproof clothing that still bears his name to this day. It was a breakthrough and waterproof Macintoshes were soon being distributed all over the world. Within a year rubber production worldwide had reached a modest 100 tons, although this would increase to 1,000 tons within six years. Finally, the rubber industry was gathering pace which, luckily, coincided with the emerging Industrial Age. However, attempts to make rubber more durable by using chemical additives were proving fruitless until, in 1839, American scientist Charles Goodyear (1800–60) accidentally discovered a reliable way to improve the properties of the natural resource.

Goodyear was born in New Haven, Connecticut, the son of Amasa Goodyear, the inventor of a lightweight hay-pitchfork that transformed harvesting methods in the old colony. He was also the first to manufacture pearl buttons and his factory supplied steel fasteners to the US Army during the War of 1812, making a small fortune in the process. In 1814 Charles was sent to Philadelphia to learn the hardware trade before returning and forming a partnership with his father in 1821. Steadily the business grew. They produced a variety of redesigned agricultural tools and by the age of twenty-four Charles was the head of a successful hardware manufacturing and supply company, which looked set to make him a wealthy man. However, in 1829, Charles's health began to suffer and a debilitating stomach illness led to the failure of a number of businesses that were linked to him. Soon the family were broke, although, as Goodyear's health began to improve, he started reading newspaper reports about the new gum-elastic product that was being introduced by the Roxbury Rubber Company in Boston, Massachusetts. Roxbury had been experimenting for some time and was trying to further improve Charles Macintosh's designs.

Intrigued by this idea, Charles Goodyear visited New York to see some of the products for himself and immediately realized that the rubber life jackets being sold were badly made and ineffective. Returning home to Philadelphia, Goodyear began experimenting with rubber products and made some robust, inflatable tubes (lifebelts) that he then presented to the manager of the Roxbury Rubber Company. Although impressed, the manager revealed to Goodyear that his company was in financial trouble as thousands of dollars of rubber goods were being returned by retailers because they were rotting and unsaleable. Although Goodyear had seen an opportunity to produce a more robust and reliable rubber product, he was also in financial trouble by then. On his return to Philadelphia his creditors caught up with him and he soon found himself residing in the debtor's prison. He was forced to sell his furniture and any other possessions and place his wife and children in lodgings. Despite this, he retained the support of his wife, who would visit him with supplies of Indian rubber, with which he would relentlessly experiment in an attempt to create a reliable material.

Eventually, by heating it, adding magnesia and moulding it by hand, Goodyear was able to make a pair of shoes and, with the help and kindness of friends, he was then able to pay off his debts and secure his freedom. Back in the heart of his family, Goodyear worked day and night in his quest to create the perfect, robust rubber product, although he eventually lost the support of his friends and creditors who, one by one, concluded that there was no commercial future for rubber. Charles Goodyear, however, had become obsessed and, once again with the support of his ever-loyal wife, travelled to New York where he set up a mini laboratory in an attic, with the help of a friendly chemist. As his efforts improved he began the three-mile daily walk to a mill at Greenwich Village where he tested various ideas and, again, his efforts fell short. Undeterred, Goodyear persuaded another friend to form a partnership and they opened a factory making waterproof clothing, rubber shoes, an early prototype life jacket and various other products. This time he achieved a modest success and was able to send for his family and buy a home for them. Sadly, once again fate intervened and the stock market panic of 1837 dissolved all of his efforts; the Goodyear family were left both penniless and homeless for the third time in a decade.

Long past the point when most inventors would have given up, Goodyear pressed on and moved his family to Boston where he reacquainted himself with J. Haskins of the Roxbury Rubber Company, who still believed in the value of rubber products. He was almost the only one and he backed Goodyear, leant him money and introduced him to all the right people in town. Goodyear soon received a patent for his rubber shoes, which he sold to the Providence Company of Rhode Island, but he was still trying to find a way of producing rubber that could reliably withstand hot and cold temperatures and acids, and avoid decomposition. His experiments would continue without success for over another three years until one day, while experimenting with rubber solutions to which he had added powdered sulphur, Goodyear spilt one of his samples onto a red-hot surface and it immediately carbonized. To some this would have been a scraping up and throwing away job. But Goodyear, however, examined his spilt sample and noticed how some of the material now had elastic properties. He had discovered a process that we now know to be vulcanization, which, to Goodyear, meant he could create either soft, flexible and waterproof rubber or a hard, durable material that could be used in many ways throughout industry, all from the same raw ingredient.

Charles Goodyear was certain he had found the answer but, because of the failure of his previous ventures, his friends and potential investors were sceptical, and for a number of years the inventor and his family continued to live in poverty. Meanwhile, he refined his methods until the day he travelled, once again, to New York City where he showed vulcanized rubber to the brothers William and Emory Ryder, who immediately recognized the value of his invention and agreed to start manufacturing various new products. Sadly, Goodyear's bad luck had followed him and the Ryder brothers' business failed, sending Charles back to the poverty house with his hopes and dreams.

Luckily, Charles's brother-in-law, the wealthy wool-mill owner William de Forest, stepped in after Goodyear's seventeen-year-old daughter Ellen demonstrated the vulcanization process to him and his business associates, the Lewis brothers, during the summer of 1843. They immediately obtained a licence from Charles Goodyear and in September of the same year the Samuel J. Lewis Company of Naugatuck, Connecticut, began manufacturing vulcanized rubber over-shoes. By 1848 four other companies were producing rubber boots under a

Goodyear licence. It was the beginning of an international rubber industry that, for 150 years, would have Naugatuck at its heart. At one point its factories had over 8,000 people working for them.

Sadly, Charles Goodyear would die before the invention of the motor vehicle, which would use vulcanized rubber tyres throughout the world, generating billions of dollars in revenue in the process. In selling his patents he failed to benefit financially from his invention and remained pragmatic about his fate, writing just before his death: 'In reflecting upon the past, as relates to these branches of industry, the writer is not disposed to repine, and say that he has planted and others have gathered the fruits. The advantages of a career in life should not be estimated exclusively by the standard of dollars and cents, as is too often done. Man has just cause for regret when he sows and no one reaps.'

However, instead of becoming a forgotten man, Charles Goodyear's legacy would remain intact for any student of invention or industry, thanks to a certain Frank Seiberling, who honoured his persistence by naming his newly formed business the Goodyear Tire and Rubber Company in 1898, just in time to capitalize on the car and motorcycle developments that would make him his fortune. Today the Goodyear Tire Company remains a multi-billion-dollar industry, and Charles Goodyear's vulcanized rubber is successfully used in millions of products around the world.

'I think there is a world market for maybe five computers.'
Thomas Watson, chairman of IBM, 1943

BLACK & DECKER WORKMATE – 'YOU WILL SELL ABOUT A DOZEN OF THOSE'

South African-born Ronald Price Hickman (1932–2011) emigrated to London during the 1950s where he spent his early years in England working for a music publisher. He soon picked up a job as a stylist with the Ford Motor Company in Dagenham before being poached by Colin Chapman's Lotus Engineering, where he became a director and was responsible for designing the first Lotus Europa car, part of the company's GT40 Project during the 1960s. Hickman was also responsible for the iconic Lotus Elan that was first introduced, to great acclaim, in 1962. Despite his success, however, Ron Hickman had other ideas and was already working on an invention that would change the building trade and the lives of DIY enthusiasts around the world.

One morning, in early 1961, Hickman was at home making a wardrobe and, lacking a workbench, he had balanced a wooden panel on the seats of a pair of Windsor chairs and placed his foot on the board to keep it steady. Then, whilst concentrating on sawing a straight line, he managed to cut through one of the seats, destroying the chair completely. The irony of ruining one piece of furniture in order to create another was not lost on the designer and he set about finding a better solution. Hickman drew a design for a portable workbench that could grip a piece of wood, had a foot platform to hold it steady and could be folded away after it had been used. In short, it could be used by any tradesman who was working alone. There and then, Hickman had designed the perfect worker's mate that required no wages, no tea breaks and was never late.

Convinced there would be huge interest in his design, Hickman approached potential investors and tool manufacturers and was surprised when, one after the other, they all rejected the idea. Giant tool-maker Stanley told him that they expected sales would be measured in the 'dozens rather than in the hundreds'. And they were not the only ones. American power-tools giant Black & Decker also rejected Hickman's workmate after telling him they did not believe the average DIY enthusiast would want to carry round such a large tool. Spear & Jackson, Salmen's and Wilkinson also turned down the opportunity to develop the workmate. Undeterred, Hickman changed his tactics and decided to take his workmate directly to the trade when he persuaded a DIY magazine to allow him to display the workmate in a corner of their exhibition stand at the 1968 Ideal Homes Exhibition in London. Within a year Hickman had sold 1,800 workmates and finally, in 1971, a new generation of product executives at Black & Decker changed their minds and decided there would, after all, possibly be a worldwide market for Hickman's workmate.

In 1972 Black & Decker began mass production of the workmate and the success of Hickman's design was instant. Within four years, one million had been sold, and by the end of the decade over ten million Black & Decker Workmates were being used by builders and DIY enthusiasts across the globe. In later television documentaries skilled craftsmen were quoted as saying things along the lines of, 'The Black & Decker is a big part of my working day and I have several in my workshop. Not having a workmate, for a builder, is like a chef not having an oven. I have no idea what people did before it was invented.' Black & Decker's Lyndsey Culf noted; 'The story of Ron

Hickman is an excellent case in point. Just because an idea is not instantly popular does not mean it is a bad idea. His effect on the world of DIY is immeasurable.' In fact, such was the success of the workmate that Ron Hickman was forced to repeatedly sue companies all over the world who had infringed his copyright. He won all of the lawsuits.

How Wrong Can You Be?
Cigar maker F. G. Alton advised John Player, in 1870, not to buy a cigarette-making factory: 'Your cigarettes will never become popular.'

In 1977 Ron Hickman re-settled his family to Saint Brélade in Jersey, where he established a design workshop and continued his association with Lotus Engineering. Former Formula One driver Derek Warwick became a close friend and remembered Hickman as a 'unique character with a distinctly inventive spirit'. Warwick also revealed that the designer, 'Came into my offices a few times with new designs and they always intrigued me. He always thought of clever ways of doing things. Whatever he saw he wanted to reinvent it. That was his mind.' Ron Hickman retired in 1982 and received an OBE in 1994 for services to industrial innovation. At the time of his death in 2011, the workmate, which a wise man had once predicted sales 'in the dozens', had shifted over 100 million units worldwide. Derek Warwick later provided the perfect epitaph when he described Hickman as: 'A little bit off the wall, a little bit eccentric and a little bit wacky. He was one of those crazy scientist types. They are not like normal people, they are different people. You don't design the Lotus Elan and the Black & Decker Workmate unless you are a very clever man.' And I think we would all agree with that sentiment.

THE BARCODE

The barcode has quite possibly been one of the most innovative inventions of modern times and has changed the retail industry more comprehensively than any other industrial event during the last century. As long ago as 1932 a group of Harvard Business School graduates began a project they hoped would simplify catalogue shopping by producing a punch card for each product listed that could be scanned, identifying each one individually. This innovative idea was initially well received but the scanners required were expensive. At that time the world was experiencing a recession on a scale never previously seen before as the Great Depression was destroying business across the globe. Retailers simply were not prepared to invest in expensive stock-control innovations. The team needed to revise their idea and make it more affordable, although the saving in labour costs their scanner method offered did appeal to some. Others argued that such modern inventory controls would also reduce warehouse injuries. Even so, it would be another sixteen years before anybody revisited the idea.

Then, in 1948, Bernard Silver (1924–63), a recent graduate of the Drexel Institute of Technology in Philadelphia, overheard a conversation between the president of Food Fair, a local chain, and one of the deans at the institute about developing a system of automatically identifying products at the point of sale, the checkout.

Silver immediately told his friend Norman Woodland (1921–2012) of the conversation and the pair began experimenting with ultraviolet ink. Unfortunately, the tests proved to be unsuccessful as the ink faded in sunlight and was far too expensive to experiment with, let alone to use on a wider scale. Their theories were rejected but, undeterred, Woodland left Drexel, moved into his father's apartment in Florida and began to work on new designs. Then one day, whilst considering Morse code as a 'language', Woodland began to draw out his first barcode system in the sand on a beach, using a piece of driftwood. As Woodland said: 'I just extended the dots and dashes of Morse code downwards and made narrow lines and wide lines out of them.' He then used a 500-watt light bulb that shone through paper and projected some of his patterns onto a 'reader'. Before long Woodland had arranged his coding system into a circle as he thought it would be easier to scan from any direction.

On 20 October 1949 Silver and Woodland applied for a patent for their 'Classifying Apparatus and Method', in which they described both straight-line and bullseye printing patterns and outlined the mechanical and electrical equipment needed to read each individual code. Their patent was approved on 7 October 1952 (US Patent 2,612,994) and the young inventors thought they were on their way. However, in 1951 Woodland had taken a job at IBM, and as soon as his patent was awarded he revealed his designs to company bosses. They were flatly rejected and Woodland then spent the next ten years attempting to persuade his senior managers to change their minds. In 1955 the US Chamber of Commerce met to consider which technologies would be advanced over the next twenty years and one of the topics featured was an electronic checkout system that would improve stock control and point-of-sale methods. Even so, IBM remained unimpressed.

Eventually, in 1961, IBM did change their minds and commissioned a study into the idea, which concluded that Woodland and Silver's designs were both practical and interesting, but also noted that the technology needed to reliably process their codes was yet to be developed. Despite this further setback IBM offered to buy the patent but, by this time, Woodland had also attracted the interest of the electronics giant Philco, and so he sold it to them instead in 1962.

Meanwhile, an undergraduate called David Collins, who was working for the Pennsylvania Railroad Company, was considering the growing need to automatically identify railway wagons. In 1959 he began work at Sylvania Electrical Products and had soon developed a system that he called KarTrak, which used red and blue reflective stripes attached to the outside of the wagons. They were coded with a ten-digit number that was unique to the company that owned the wagons and that also identified each individual wagon. The light that then reflected off each unique set of stripes was fed into a photomultiplier that was able to distinguish between red and blue. Collins's invention was tested by the Boston and Maine Railroad between 1961 and 1967 on their gravel cars, and the Association of American Railroads began to install the KarTrak identifying system throughout the entire fleet of railroad cars on 10 October 1967. Unfortunately another economic recession led to a number of bankruptcies throughout the industry during the early 1970s. This, coupled with the fact that dust and dirt led to unreliability, meant the KarTrak installation was abandoned by 1978.

Although the railroad experiment had failed, a toll bridge in New Jersey commissioned the system as a way of scanning cars that displayed a monthly crossing permit, and soon afterwards the US Post Office employed a similar technique to monitor trucks moving in and out of its depots. Then the pet-food manufacturer Kal Kan commissioned Sylvania to supply a cheaper version to simplify the tracking of its warehouse and distribution facilities. Finally the sleeping giant IBM fully embraced the idea and commissioned a team to develop a barcode system that could be printed on all products, and by 1973 it had already convinced many grocery manufacturers and retailers that they would benefit from the new scanning method of identifying all of their products. Developers had predicted that 75 per cent of all consumer products could be labelled by 1975, although two years later only 200 retail outlets actually had scanning machines installed at their checkouts.

It was obvious that the only way that a universal scanning system would be effective was if the majority of retailers installed scanners. Until they did, suppliers would be reluctant to spend money on allocating and then printing unique barcodes on their packaging. It was a stalemate between retailers and suppliers, with neither side wanting to commit their hard-earned resources before the other did so. Why would a retailer buy expensive scanning machines if the barcode method failed to be rolled out to all products? And why would suppliers print barcodes if the retailers had no means of reading them? *Businessweek* magazine was soon dismissing the entire initiative as 'the supermarket scanner that failed'.

However, news was coming in that retailers who had installed scanning machines were reporting sales increases of 10–12 per cent within the first six weeks, and that those higher sales levels continued. Suppliers were soon able to demonstrate that the return on investment for each installed scanner was over 40 per cent, and by the end of the 1970s over 8,000 stores were converting to the scanning method of sales checkout. But globally, this was not enough. And there was opposition and resentment from the unlikeliest of places. Conspiracy theorists dismissed the scanning technology as an intrusive means of surveillance and some Christian groups were loudly complaining that each code secretly used the number 666, the number of the devil. Some television presenters were warning of barcodes being a 'corporate plot against ordinary consumers'. Happily, all such ill-considered opposition was eventually overcome and barcode systems were refined, updated and then rolled out across the world. They were soon printed on just about every product imaginable, including medicines, airline tickets and patient wristbands. Bernard Silver and Norman Woodland's initial idea, in 1948, had overcome hurdle after obstacle and, after half a century, finally become part of our everyday lives. And it is the sole reason, in today's busy retail outlets, that we don't spend most of each weekend standing in a queue.

Airplanes are interesting toys but of no military value.
Marshal Ferdinand Foch, Professor of Strategy,
École Supérieure de Guerre, in 1904

THE TIN CAN

At the same time as Nicolás Appert (1749–1841) was answering the call of Napoleon Bonaparte and devising vacuum-sealed bottles to preserve food for the French Army (see 'Why Chicken Kiev Rules Supreme'), ironically it was a British merchant called Peter Durand who was awarded a patent for the tin can in 1810. As England was at war with France (as usual), the little Frenchman did not get hold of the designs.

Very soon afterwards, in 1813, John Hall opened the first commercial canning factory in the world, although output was slow and tedious, with only a half a dozen cans being produced every hour. It would be another thirty-three years before Henry Evans invented a machine that would increase output to around sixty per hour. However, these early tin cans were impractical and unpopular. For a start they were thick and heavy (heavier than their contents), and they had to be hammered open. The instructions for users were to employ a hammer and chisel and therefore, needless to say, canning remained a long way behind bottling.

This was until 1858, when cans became thinner and it was possible to produce a tool specifically for opening them. The first can opener was patented in 1858 by Erza Warner of Waterbury in Connecticut, and mass production began just in time for the American Civil War (1861–5). By then, soldiers rations were regularly delivered in cans, and by the later years of the war the famed pickled 'bully beef' was in regular supply. In 1865, just as the war was coming to an end, the 'bull's head' opener had been invented and distributed throughout the ranks. A full fifty years after the first tin can had been invented, along came the can opener.

These were quickly developed, and within a year J. Osterhoudt had patented a tin can with a key opener attached that could remove a thin band of tin from around the top to allow easy access to the contents. Then in 1870 William Lyman designed the classic can opener that used a wheel to roll and cut the top of a can away. But it would be another half a century before the Star Can Company of San Francisco added a serrated edge to one of the wheels, called a feed wheel, which would grip the edge of a can whilst cutting it. This is the design that is still in use and popular today. Incredibly, aside from the electric version, nobody has been able to improve on that basic design. In all, it took over a century since the tin can had been invented to mass produce a practical tin-can opener.

> We have now reached the limits of what is possible for computers.
>
> John von Neumann in 1949

GOING UNDERGROUND – YOU WANT TO RUN STEAM TRAINS WHERE?

The first over-ground railway line to be completed in London was the Greenwich to London Bridge line that opened in 1836, although it was originally proposed by a former Royal Engineer, Colonel George Landmann, as early as 1824. It took an Act of Parliament in 1833 to finally approve the building of the railway that was intended to eventually run all the way to Dover, England's main gateway to Europe. Even though Robert Stephenson's *Rocket* steam engine was only demonstrated in 1829, the whole idea of a national rail network had been discussed since the 1700s, when horse-drawn trams guided by rails had become a regular feature in the main cities of England. It was the steam engine that provoked more ambitious plans. In fact, plans for the rail network were so ambitious that the first proposal for an underground rail line was submitted in 1830, six years before the over-ground had been completed. As the intercity network grew, so did the plans for an underground system that would primarily link the main London terminals, although Parliament resisted and opposition was fierce. The idea of running steam trains full of passengers through enclosed tunnels, deep under the streets of London, was simply out of the question.

For the best part of twenty years the debate rumbled on and the over-ground network grew ever busier and more successful. A great migration was taking place and thriving communities were growing up around stations that were up to an hour's train ride from London, as city workers became commuters. And still the London authorities refused to connect the terminals to the City of London, Westminster, Fleet Street or each other via underground rail lines. Over-ground trains were not permitted to expand any further into the capital than where they already terminated, and this meant nearly one million new commuters were arriving in the city each morning with no means of reaching their places of work, other than by foot. By the 1850s London was not only the world's largest industrial city, it was also the financial centre of global trading and one of the busiest ports in Europe. This huge increase in traffic naturally led to a complete gridlock, with horse-drawn vehicles, commuters, hand carts, imports, exports, men, women, children and animals all competing for the same space on the same narrow streets. It needed an ambitious solution to link the city terminals to the centre of London and eventually there was only one way to go. They were going underground.

The driving force behind the first underground line was the radical MP for Lambeth and Solicitor to the City of London, Charles Pearson (1793–1862), who campaigned for a line to be built between Paddington station and the City of London. Pearson was known for campaigning on behalf of women's rights (they didn't have very many in the mid-nineteenth century), the abolition of the death penalty, penal reform and in favour of stiff penalties for corruption when it came to jury selection. A man of the people, he used his position as Solicitor to the City of London to call for improvements in transport communications, particularly by tunnel.

In 1845 Pearson published a pamphlet outlining the benefits of a tunnel link to the financial district using an atmospheric railway that would harness compressed air to push carriages through the tunnel. His plans were rejected and his proposals ridiculed. The following year he enlisted the support of the City of London Corporation and proposed a direct rail link to Farringdon, although, once again, the Royal Commission on Metropolitan Railway Termini threw out his plans. In 1856 another Royal Commission was set up to report on London's growing congestion problem and Pearson once again joined the debate. He wrote: 'The overcrowding of the city is caused, first by the natural increase in the population and area of the surrounding district; secondly, by the influx of provincial passengers by the great railways north of London. And the obstruction experienced in the streets by omnibuses and cabs coming from their distant stations, to bring the provincial travellers to and from the heart of the city. I point next to the vast increase of what I may term the migratory population, the population of the city who now oscillate between the country and the city, who leave the City of London every afternoon and return every morning.'

This time there were many proposals for underground railway lines and, again, they were all rejected. Much of the opposition to an underground railway centred on the difficulty of digging deep tunnels, especially underwater, and many previous attempts had ended in both failure and loss of life. It is worth remembering here that between 1830 and 1860 there was no electricity, and using steam power, and shoulder power, to dig deep tunnels and move huge quantities of earth to the surface was a tricky business, even for the best engineers and the hardest of grafters. But this wasn't the only problem. The main obstacle would be attracting commuters to actually use the underground network.

Most people had never even been underground, unless they were miners, and the thought of travelling at speed through underground tunnels made the blood run cold. Many Victorian novelists (the rock stars of their day) were predicting death and destruction. The horror writers were having a fabulous time and ordinary folk were scared to death at the thought of travelling beneath graves and sewers. Worse still, the national newspapers, led by the *Times*, were running fierce editorial campaigns against underground rail travel and were predicting passengers soaked in dripping sewage in tunnels infested with rats and full of poisonous fumes seeping from gas mains. There were warnings of deafening noise, steam, smoke and suffocation. One famous editorial, in 1861, noted that it would be 'insane to imagine that horse-drawn omnibus passengers would instead choose to be driven amid palpable darkness through the foul subsoil of London'. They went on to predict that the very notion of the underground railway would in the future be associated with other absurdities such as plans for a tunnel under the English Channel. Others were suggesting homeless families would move into the tunnels and whole communities would form, spawning albino children who would never see the light of day and grow up with 'guttural tongues'.

But the problem above the surface was growing worse and London was becoming slower and even more congested. Finally, Parliament stepped in and a Private Member's Bill for the Metropolitan railway link between Paddington and Farringdon was passed on 7 August 1854, just as Pearson had proposed nine years earlier. Although not directly involved, Pearson used his influence in the City to raise £1 million for the construction of the line and wrote more pamphlets promoting the project. A charitable act given that he would never need to use the line as he lived in Wandsworth, south of the City. By 1860 the money had been raised and work began on clearing slums and carving a tunnel beneath London's busiest streets. Unfortunately, Pearson would not live to see the completion of the world's first underground railway as he died of dropsy (excess fluid) on 14 September 1862, four months before the opening of the original Metropolitan Line.

In its first year of operation over nine million passengers used the underground railway that only two years previously the influential *Times* had confidently predicted 'nobody would ever use'. During the next fifty years the

'London Underground' grew to become the largest and most profitable underground railway network in the world and influenced similar projects in hundreds of other overcrowded cities.

How Wrong Can You Be?

David Riesman, an American social scientist, took his life in his hands during 1967 when he uttered the words, 'If anything will remain more or less unchanged it will be the role of women.'

ACCIDENTAL INVENTIONS

SACCHARIN

What is small, pink and can be seen almost everywhere you go? Certainly anywhere serving coffee or tea. In 1879 chemist Constantin Fahlberg (1850–1910) was working at the Johns Hopkins University analysing the chemical properties of coal tar, a by-product of coal. His brief was to find alternative and profitable uses for the natural resource. After one particularly long day in the lab Fahlberg returned home to find his wife had baked a tray of his favourite biscuits and so the hungry chemist tucked in. The first thing he noticed was that his biscuits tasted a lot sweeter than usual, and he asked his wife how much sugar had been added. After she had assured him that exactly the same recipe had been followed as usual, Fahlberg realized he had failed to wash his hands after work. Tasting his fingers, he immediately understood that one of the by-products of coal tar was that it was a natural sweetener. Fahlberg gave his discovery the name 'saccharin', which means 'relating to' or 'resembling' sugar. Although it was soon commercially available as a calorie-free sugar alternative, it would be the sugar shortages during the First World War that ushered in worldwide distribution of Fahlberg's accidental discovery.

POPSICLES

Frank Epperson (1894–1983) of Oakland, California, was only eleven years old when, in 1905, he began experimenting with soda powder, fruit flavouring and water in an effort to invent his own soda drink. This wasn't unusual as soda and water was a popular mix in those days and folk were used to making their own potions. It was a cold winter evening when Epperson was called inside and he left his experiments on the porch. During the night temperatures fell below zero and, the following morning, Epperson found his fruit-flavoured soda drinks had all frozen solid, with the stirring stick left in each one. Apparently he tasted them, showed his friends and allowed them to sample his frozen specimen. And then he forgot all about his accidental invention for a number of years.

Then, in 1922, Frank was asked to contribute to the local firemen's fundraising ball and he went along with as many of his popsicles as he could carry. They went down a storm and he was sold out within minutes so, sensing an opportunity, Epperson applied for a patent for his 'frozen ice on a stick'. He had called his invention an 'Eppsicle' and he began producing them in many different flavours as a treat for his own children. The kids, who called him Pop, then started calling them a Popsicle, the name under which they would become internationally famous. Two years later, in 1925, Epperson sold the rights for his Popsicle to the Joe Lowe Company in New York, who began distributing the Popsicle and selling them to children for a nickel each.

Before long, the brand had expanded to include the Fudgsicle, Dreamsicle and Creamsicle, and in 1965 Consolidated Foods Corporation acquired the rights.

217

Today hundreds of millions of Popsicles are bought in the United States each year and are available in over thirty flavours. Many rival manufacturers have produced variations of the Popsicle but, it is fair to say, whatever form of frozen iced lolly you have eaten, you can trace its origin to the accidental invention by an eleven-year-old boy called Frank on a cold winter's night over a century ago.

VIAGRA

During the 1990s, down at a chemical research facility in Sandwich, Kent (see 'Did an Old Card Player Invent the Sandwich?'), where work was being undertaken for the American pharmaceutical giant Pfizer, a group of British scientists were researching a substance called sildenafil citrate. It was hoped that they would develop a new and more effective way of controlling high blood pressure, angina and altitude sickness. Pfizer had expected the drug to relieve debilitating chest pains caused by angina, and the early samples were sent for testing at the Morriston Hospital in Swansea under the supervision of Ian Osterloh. On conclusion of the tests Osterloh noted that the drug had little effect in relieving angina, although he did observe a few adverse side effects including headache, hot flushes, congestion, blurred vision and vigorous erection. Wait a minute, '*adverse* side effects'? Pfizer were quick to notice the potential of this new drug and filed for a patent in 1996, receiving approval for its use in treating erectile dysfunction by the Food and Drug Administration of the United States on 27 March 1998.

Viagra, the trade name given to the drug, was soon being marketed to Americans after being famously endorsed by US Senator Bob Dole and Brazilian soccer

218

legend Pelé. Bizarre choices in the extreme, but the campaign was a huge success, with sales of Viagra soon reaching the $2 billion-a-year mark. Other uses have been found for Viagra, although we can only imagine how these conclusions were reached or who commissioned the tests. These include an Argentine university discovering that Viagra reduces jet lag in hamsters; professional athletes believing the expanding of their blood vessels will enhance performance; and separate studies in both Israel and Australia concluding that 1 mg of Viagra dissolved in a vase of water will increase the shelf life of cut flowers. I would love to have been in the room when scientists announced that particular discovery. I would want to know what they *thought* they were going to find out.

COCA-COLA

John Stith Pemberton (1831–88) was born in Knoxville, Georgia, and was the nephew of legendary Confederate General John Clifford Pemberton, a veteran of the Frontier Wars. As a teenager the younger Pemberton was enrolled in the Reform Medical College of Georgia in Macon and graduated, at the age of nineteen, with a licence to practise pharmacy. By April 1865, as the American Civil War was coming to a close, John Pemberton had followed his famous uncle into the Confederate Army and was serving with the 12th Cavalry when he was wounded at the Battle of Columbus, Georgia. In close-quarter combat Pemberton had been sliced across the chest with a sabre and, as was common with many wounded soldiers of the time, became addicted to the morphine he was given to ease the pain. As a trained pharmacist Pemberton began experimenting with opium-free alternatives in a bid to cure his addiction, and his first effort, Dr Tuggle's Compound Syrup of Globe Flower, soon became a popular and successful cough remedy.

Pemberton's next experiment, with the coca plant and kola nut mixture, became Pemberton's French Wine Cola and was marketed to the intellectual classes. Pemberton himself described the drink as a 'remedy that would benefit physicians, lawyers, scholars, poets, scientists and divines'. Pemberton claimed his remedy would help anybody 'devoted to extreme mental exertion'. In fact, all Pemberton's recipe seems to have been was cocaine mixed with French wine, which had, in fact, first been introduced by a Parisian chemist, Angelo Mariani, in 1863. Mariani's Vin Mariani had become instantly popular among the great and the good of society, and authors Arthur Conan Doyle, Alexandre Dumas and Jules Verne were all known to be big fans. Apparently, Pope Leo XIII carried a flask everywhere he went and even awarded Mariani a medal for his invention. In America Pemberton made many outrageous claims for his version of Mariani's concoction, including that it was a 'wonderful invigorator of the sexual organs'.

In 1885 Atlanta and Fulton County introduced a temperance law and Pemberton was moved to develop a non-alcoholic version of his popular 'remedy'. Luckily for him, the new law did not restrict the use of the coca ingredient and his non-alcoholic cocaine remedy, unsurprisingly, became very popular. Unfortunately, as a remedy it did not work so well for its creator, and Pemberton fell ill. On the verge of bankruptcy, he initially wanted to retain the rights to Coca-Cola for his son Charles, but he was also on the verge of bankruptcy himself and persuaded his father to sell out to Asa Griggs Candler for only $550 in 1888. (Later, historians would suggest that Pemberton's signature on the sales deed was a forgery, probably by Charles himself.) At the age of fifty-seven, John Pemberton, who was sick, desperate, still addicted to morphine and suffering from stomach cancer, died in August of that year. His son Charles, who continued to sell an alternative to his father's invention, died six years later, himself addicted to opium.

And so it was Candler, a devoted teetotal Methodist, who turned Coca-Cola into a nationally popular soda drink and made himself a multi-millionaire in the process. Today, thanks to the Temperance Movement of Atlanta and Fulton County, Coca-Cola, John Pemberton's accidental medicine, distributes nearly two billion servings to over 200 countries worldwide, every day.

> There is no chance that the iPhone is going to get any significant market share. No chance.
>
> Steve Ballmer, Chief Executive Officer
> of Microsoft, 2007

THE HAMBURGER: FROM A GERMAN SNACK TO AMERICAN ICON

The hamburger is commonly acknowledged as America's signature dish and it has been estimated that over fourteen billion of them are eaten there every year. So you can see how being the home of the hamburger could mean big business, and this is indeed the claim of the small town of Seymour in Wisconsin. Seymour's Hamburger Hall of Fame celebrates the history of burgers, while the town's annual one-day Burgerfest includes parades and competitive events themed on burger-related condiments, including the famous Ketchup Slide. In 1989, the world's largest hamburger was produced during the festival, weighing in at a hefty 5,500 pounds (nearly 3 tons). Back in 1885, as the town informs its visitors, fifteen-year-old Charles Nagreen (1870–1951) was selling minced-meat patties from his stall at the very first Seymour Fair. Hamburger Charlie, as he became known, soon worked out that his meatballs weren't selling because customers were unable to easily walk around eating them by hand. So Charlie flattened his meatballs, served them between two slices of bread and called them hamburgers, after the hamburger steaks that were already popular in the northern states.

Charlie Nagreen continued to sell hamburgers at county fairs for the rest of his life and became a local celebrity, claiming right up until his death to be the inventor of this form of fast food. But there is a rival claim, also dating from 1885. According to the other story, sausage-making brothers Frank and Charles Menches were accidentally sent beef instead of pork by their supplier. Short of time and with limited resources, they decided to cook up the beef instead and serve it in sandwiches at the Erie County Fair, naming their new concoction the hamburger sandwich after their home town of Hamburg in the state of New York.

And there are all kinds of other counter-claims: no less an institution than the US Library of Congress has credited Louis' Lunch, a restaurant in New Haven, Connecticut, with making America's first burger in 1895. But eminent hamburger historians (and yes, they do exist) have begged to differ, citing instead Old Dave's Hamburger Stand at the 1904 World Fair in St Louis, run by Fletcher Davies, a restaurant owner from Athens in

Texas, as the original inventor. So convincing was their argument, in fact, that the Texas state legislature then confirmed Athens as the 'Original Home of the Hamburger' in November 2006. They take this all very seriously over in America.

SAFETY GLASS

In 1903 French artist and scientist Édouard Bénédictus (1879–1930) was working alone in his studio when he climbed a stepladder to reach something from a top shelf. As he fumbled around, Bénédictus accidentally knocked a heavy glass flask and was too slow to prevent it falling to the floor. There was an almighty crash, and when he climbed down to survey the mess he was amazed to find the flask, although shattered, remained intact. In fact, it had hardly changed shape. Bénédictus had never seen anything like it and decided to investigate. He soon realized that, although the flask was empty, it had previously been filled with a solution of cellulose nitrate, a transparent liquid plastic, which had been poured away but had left a thin, clear film coating inside the glass. Apparently, an assistant had been too lazy to wash it properly and returned it directly to the shelf instead.

To begin with, the scientist thought very little of his discovery. Then, one morning later in the same week, Bénédictus was reading the newspaper when he came across a feature about the new fashion of automobiles and, more to the point, the series of collisions that had occurred between Parisian drivers. Most of them, he read, had been seriously injured by shattered windscreens and flying glass. He later recorded in his personal diary: 'Suddenly there appeared before my eyes an image of the broken flask. I leapt up, dashed to my laboratory and concentrated on the practical possibilities of my idea.'

For the following twenty-four hours Bénédictus experimented by coating glass objects with layers of liquid plastic and then smashing them. He then recorded: 'By the following evening I had produced my first piece of Triplex, which is full of promise for the future.' But there was resistance from the automobile industry, who rejected any ideas for safety windscreens as they were already struggling to keep their costs down on what was, at that time, an expensive and unnecessary luxury item. The general attitude of the day was that the driver, and not the automobile designer, was responsible for road safety. Automobile designers simply were not interested in avoiding, or minimizing, injury in the event of an accident.

Bénédictus would have to wait another decade and for the outbreak of the First World War before his invention would be incorporated into the production of gas masks. Military designers found it relatively easy to coat small oval lenses with a plastic solution, which provided the sort of protection that was desperately needed at the time. Even the transporting of gas masks was likely to result in a high percentage of breakages before they even reached the front lines, without Bénédictus's accidental invention. It was only after car makers had learned of the benefits of safety glass under battle conditions that they finally began to incorporate the new technology into their own designs.

There is growing evidence that smoking has pharmacological effects that are of real value to smokers.
                                President of Philip Morris in 1962

PENICILLIN

The invention of penicillin is, quite possibly, the most famous accidental invention of all time. It has been a story told and learned by schoolchildren for generations, although, just in case you missed it, we shall repeat it here.

Alexander Fleming (1881–1955) was a British botanist, pharmacologist and biologist who worked at a London shipping office for four years until an inheritance from an uncle enabled him to enrol, in 1903, at the St Mary's Hospital Medical School in Paddington. At the age of twenty-one, Fleming had no particular interest in science or medicine, although his elder brother was already a physician and strongly urged his younger sibling to use the money wisely and study for a professional career. Fleming earned his degree in 1906 and graduated with distinction. During medical school he had been a leading member of the shooting team of St Mary's and the club captain, eager to keep him on the team, recommended him to the research department at the medical school. There, he became the assistant to Sir Almroth Wright, a pioneer in the field of immunology and vaccine research. At the outbreak of war in 1914, Fleming joined the Army and served as a captain in the Royal Army Medical Corp. During this time the world read stories of soldiers from all sides becoming victims of the new automatic machine guns, explosive artillery fire and mustard gas.

But the young field doctor began to notice something far more dangerous than modern weapons of war. Fleming realized that most fatalities were caused by infection from minor wounds that were being treated in field hospitals all along the western front. At that time the primary prescription for open wounds was a liberal supply of cheap antiseptic, and it became obvious to Fleming that this was possibly more dangerous than applying no treatment at all. He was unwilling to accept the inefficiency of modern medicine and vowed to dedicate his career to identifying, understanding and fighting infections. He was particularly motivated to find a safer treatment than what he considered to be 'deadly' antiseptic. After the war he returned to St Mary's Hospital and, with Sir Almroth Wright's encouragement, he studied antiseptics and their unintended effects, making a major discovery in 1923 when he identified the enzyme lysozyme in human mucus. Fleming observed how this naturally occurring agent protected the human immune system from certain bacteria.

By 1928 Alexander Fleming was leading a research team in a study of common bacteria that was spreading disease through urban areas. As Professor of Bacteriology at the University of London, Fleming could have been expected to set an example but, in fact, it was his untidiness that would change the world of medicine, modernize the human effort against disease and save millions of lives. In August that year, the Professor went on holiday with his family and, before he left, stacked all his equipment, including Petri dishes, into a corner of his untidy laboratory. On 3 September, Fleming famously returned from his holiday and, as he was setting out his equipment, noticed that he had failed to clean them properly before he had left. As a result of this he observed that one of his samples was contaminated with a fungus and that the bacteria immediately around it had been destroyed. He was about to throw it away when he showed it to his former assistant, Merlin Price, who reminded him, 'This is how you discovered lysozyme.' Over the following weeks Fleming began to experiment with the mould and discovered he could easily produce a substance that naturally killed any number of harmful bacteria, many of which caused disease. Fleming later recalled: 'When I woke up just after dawn on 28 September 1928, I certainly did not plan to revolutionize all medicine by discovering the world's first antibiotic, or bacteria killer. But, I suppose, that's exactly what I did do.'

Fleming identified his discovery as part of the penicillium genus and, after many months of calling it mould juice, released a paper describing 'penicillin' on 7 March 1929. Two other scientists, a Nazi refugee called Ernst Chain (1906–79) and Australian Howard Florey (1898–1968), developed Fleming's penicillin further so that it could be produced as a drug, which was immediately effective, although supplies remained limited and expensive. They would have to wait until 1940 and the start of the Second World War before American drugs companies began to mass-produce penicillin. Alexander Fleming became internationally famous for his accidental discovery and was elected a fellow of the Royal Society in 1943, knighted for his services to medicine in 1944 and, in 1945, he shared the Nobel Prize for Medicine with Florey and Chain. The laboratory where he made his life-changing discovery is preserved as the Alexander Fleming Laboratory Museum at St Mary's Hospital in Paddington, London.

There is a great story of the connection between Alexander Fleming and Britain's gritty wartime Prime Minister, Winston Churchill, which, for the record, I don't believe is true. But it is still a good story and goes something like this.

227

Hugh Fleming (1816–88) was a poor Scottish croft farmer, working the land to provide his young family with food and clothing in the first instance, but he dreamt of providing them with a better future and a better life than he himself had endured. One morning the farmer heard cries for help coming from a nearby field and he dropped his tools and ran in the direction of the voice. There, waist deep in a murky Scottish bog, was a terrified boy trapped and sinking into the ground. Thinking not of his own safety, Fleming went straight in and pulled the boy from danger, saving him from certain death.

The following day a grand carriage drew up at the modest croft cottage and a noble lord stepped out to greet the farmer. He introduced himself as the father of the boy Fleming had saved and insisted that he wanted to reward the farmer to show his sincere gratitude. But Fleming refused, declaring he had only done what any-one else would have done in the same circumstances. At that point, the farmer's own son joined his father. 'Is this your boy?' asked the nobleman, and Fleming proudly agreed that it was. 'In which case I will make you a prom-ise,' said the lord. 'I will take the boy and pay for the best education money can buy. If he is anything like his father he will grow into a man we will both be proud of.'

Seeing a chance for his son to escape a life of poverty, the farmer agreed and the boy then benefited from the finest education, eventually graduating from St Mary's Hospital Medical School in London. He was later knighted for his contribution to medicine and became known as Sir Alexander Fleming, the man who discovered penicillin. Some years later the nobleman's own son became seriously ill with pneumonia and it was the crofter's son's penicillin that saved his life, truly repaying

Lord Randolph Churchill's benevolence. His son, the boy dragged from the bog and whose life was saved for a second time by the Fleming family, was Sir Winston Churchill.

It is a common enough tale and has been circulating for many years. Unfortunately, it seems it is untrue, with Fleming himself, quoted in the book *Penicillin Man – Alexander Fleming and the Antibiotic Revolution*, dismissing the story as 'a wonderful fable'. It is known that Churchill consulted with Sir Alexander Fleming on 27 June 1946 about a staphylococcal infection, which had apparently resisted treatment by penicillin. However, there is no record of a young Churchill nearly drowning in Scotland or of Lord Randolph Churchill paying for Fleming's education. But you never know.

How Wrong Can You Be?
Between 1885 and 1891 the US Geological Survey announced that there was little or no chance of oil being discovered in California, Texas or Kansas, and the US Department of the Interior predicted, in 1939, that American oil supplies would last for only thirteen more years.

SUPPRESSED INVENTIONS: TRUE OR URBAN LEGEND?

THE RIFLE BEAM

The Rifle Beam has been presented as the most effective cure for all cancers. That's right, the cure for cancer has already been discovered. However, many believe the American Medical Association has deliberately discredited the invention and ordered a cover up. After all, there are enough people in the world already. Why start saving more lives?

THE LIGHT BULB

The original incandescent light bulb was invented by British chemist Sir Humphry Davy (1778–1829) in 1805, although it would be another seventy-five years before Thomas Edison found a way for them to be used commercially. Then, in 1924, the leading light-bulb manufacturers formed the International Phoebus Cartel with the aim of standardizing light-bulb fittings. However, many believed that the cartel was actually formed to suppress the invention of the life-long light bulb. It was a new design that would never need to be replaced. In fact, the cartel went one step further and agreed to actually limit the life expectancy of a light bulb, which would increase the demand for replacements. Some men of science claim that the patent for the lifetime light bulb, along with its technical information, is 'buried somewhere in a drawer' at the head office of one of the major light-bulb manufacturers.

It should be pointed out that there is no actual evidence for this, although it is known that the average lifespan of a light bulb, in the Western world, is around 2,000 hours, whilst those in the former communist countries, who were not part of the cartel, have around double that life expectancy. Modern Chinese bulbs are estimated to last up to three times as long. Now, despite there being no evidence to support such claims, it is known that German watchmaker Dieter Binninger (1938–1991) invented a light bulb that was estimated to last for 150,000 hours of continual use, or eighteen years. However, soon after finding a manufacturer who agreed to actually produce them, Binninger mysteriously died in an aircraft accident in 1991 and his invention quietly disappeared from focus. Murder in the murky light-bulb business, or conspiracy theory?

COLD FUSION

Cold fusion is a stable form of nuclear energy that has been safely produced at room temperature. It could be further developed in a way that would eventually provide free energy for the entire world population. It was dismissed by authorities and funding for further tests was cancelled.

THE CHRONOVISOR

This was developed with claims that it made users able to see both backwards and forwards in time. It was dismissed as a fake and disappeared, although some believe it is alive and well and being developed further in the Vatican.

WARDENCLYFFE TOWER

Wardenclyffe Tower, situated in Shoreham, New York, was the centre for experiments with wireless electricity conducted by Nikola Tesla (1856–1943). Many believed Tesla was on to something, but his funding was withdrawn after investors realized it could not be metered, and free electricity would lead to zero profits.

THE CLOUDBUSTER

The rain-making 'cloudbuster' was invented by Dr Wilhelm Reich (1897–1957) and, apparently, successfully tested in 1953. Reich was a controversial scientist who was later arrested and all of his scientific notes were destroyed.

THE ANTI-GRAVITY DEVICE

Thomas Townsend Brown (1905–85) developed an anti-gravity device by using discs that harnessed electrogravitic propulsion. Apparently the effects were so successful that the development was immediately classified as top secret by the US government. Nothing was ever heard of it again.

THE EV1

Released in 1996, the EV1 was the first electric car to be successfully mass produced. General Motors destroyed the prototypes and discontinued all other research, allegedly after pressure from the oil companies.

THE WATER FUEL CELL

The water fuel cell designed to replace petrol was invented by Stan Meyer (1940–98), but his claims were declared fraudulent by an Ohio court in 1996. Some claim the technology has been suppressed; after all, who wouldn't want free energy from a naturally unlimited resource (the sea)? But inventor Meyer would not give in easily and there are certainly suspicious circumstances surrounding his death.

THE IMPLOSION GENERATOR

This also would have supposedly provided free energy for everybody had Austrian inventor Viktor Schauberger (1885–1958) not been silenced and discredited by his business partners.

PROJECT XA

In the late twentieth century Project XA supposedly invented safer cigarettes that had harmful carcinogens removed. The powerful tobacco giants resented the implication that their product was unsafe and the XA experiment was abandoned.

> Louis Pasteur's theory of germs is ridiculous fiction.
> Pierre Pachet, Professor of Physiology
> at Toulouse University, in 1872

RIDICULOUS INVENTIONS YOU WISH YOU HAD INVENTED

SPANX

In 1996 a twenty-five-year-old graduate of Florida State University, Sara Blakely, started working for an office-supply company selling fax machines. Part of the company's dress code included ladies wearing tights (that's pantyhose if you are reading this in America), which Blakely resented in the hot sunshine of Florida as she liked to wear sandals. However, she did like the way the top section of the tights made her look slimmer and eliminated underwear lines that were visible through her outer clothing. Blakeley experimented by cutting them off just above the knee but found the material rolled up her legs as she walked around. She then spent the next two years trying out various materials before filing for a patent. Her Spanx underwear was initially rejected by every manufacturer and retail outlet she approached over the following three years, until Highland Mills agreed to a production deal after the owner's daughters both endorsed Blakely's underwear. The Spanx brand made $4 million in its first year of trading and is now estimated to be worth $1 billion.

THE SNUGGIE

A unisex body-length blanket with sleeves is how the Snuggie is usually described. The sleeved blanket started life in 1998 as a Slanket and was first marketed as the Freedom Blanket when it was displayed in shops. Student Gary Clegg's mother made him a wrap-around blanket with a single sleeve that he could wear in his cold dormitory room and still be able to work with his one free hand. Clegg later added the second sleeve to create a product that initially sold for $14.95 or $19.95 for a pair. Launched in 2008, the Snuggie variation sold four million units by the end of 2009 and is now responsible for over half a billion dollars in worldwide sales. Half a billion dollars for a back-to-front dressing gown.

THE PLASTIC WISHBONE

Who would have thought it? A fake wishbone to keep the kids quiet on Christmas Day or Thanksgiving instead of arguing over who was given the bird's single furcula bone. Lucky Break Wishbones, the 1999 brainchild of inventor Ken Ahroni, is now keeping spoiled kids happy to the tune of $2.5 million a year. Now why didn't I think of that?

THE HEADON

This is a wax product with an annoying television commercial that claims to be able to cure a headache by simply rubbing it across the forehead. Despite there being no scientific research to support these claims (and for legal reasons I am not suggesting that it doesn't work) over six million sticks of the stuff were sold in 2006 alone, at a price of $8 a time. You can work out the math.

BILLY BIG MOUTH BASS

Presumably everyone has seen one of these, or at least heard of the singing fish that became the must-have novelty during the 1990s. Who would have thought that over one million of them would be sold in the year 2000 alone, at a cost of $20 each?

THE BEANIE BABY

Nobody took the Beanie Baby seriously at all, apart from its inventor Ty Warner (b. 1944) who, apparently, sold 300,000 at his first toy show. Presumably that means orders as it is hard to believe anybody would take that many units to a product launch. Who cares anyway as Beanie Baby sales have now topped five billion and still counting. Ty Warner is estimated to be worth $3 to $6 billion.

TAMAGOCHI

Imagine going into a meeting with an idea for an electronic pet that needs constant attention to either shut it up or stop it from dying. It would take either a brave man or a fool. Or both. Until it sold seventy-four million units, making billions of dollars for everybody involved. Gin and tonics all round, I think.

IFART

An application for smartphones that replicates the sound of twenty-five forms of flatulence to please the tiniest of minds. It even has a 'record-your-own-fart' feature. Inventor Joel Comm (b. 1964) must have hoped it would prove to be popular, but he could never have predicted 114,000 downloads at $1 a time during the first two weeks of uploading it to iTunes. The iFart app went to number one on the application chart where it stayed for three weeks, making it the biggest-selling app in the world during that time. It is now thought to have received over a million downloads, earning its creator more than I care to think about.

YELLOW SMILEY FACES

In 1963 Harvey Ball (1921–2001), a designer for a PR company, was asked to think of a logo for one of their clients, a life assurance company. In no time at all he had produced the goofy, yellow, smiling cartoon face and added the words 'Have a nice day'. A few years later Bernard and Murray Spain were planning to open a novelty store and thought the smiley face would be a good logo for them to use, so they bought the rights. They then used that image on just about anything they could think of, including key rings, Frisbees and carrier bags, and were soon producing a vast range of products with the smiley face logo. By 1971 sales had reached fifty million in number and the brothers' novelty store was expanding into a chain. In the year 2000 they sold the business for a handsome half a billion dollars. The man who designed the logo in the first place was paid $45 for his efforts.

THE WACKY WALL WALKER

Whatever possessed Ken Hakuta to spend $100,000 to buy the rights to a toy that sticks to walls when it is thrown, and then appears to walk down them, is anybody's guess. His mother had sent him one whilst on a visit to China and Ken was convinced that the toy would be a big hit in America. He was wrong. At least to begin with. Sales were painfully slow but then somebody at the *Washington Post* stumbled upon one and wrote a review. The ensuing craze that followed led to the Wall Walker becoming one of the biggest fads of all time with a reported 240 million of them being sold in just a few months, earning Hakuta a healthy $80 million in the process.

THE SLINKY

Richard James (1914–74) was a naval engineer and one day, during the Second World War, he was working with a new tension spring, which he clumsily dropped. James and his colleagues then watched as the spring worked its way across the floor by using its own momentum. By the end of the war James had decided to make a toy out of it but was so nervous that he persuaded a friend to keep him company at the initial launch. Both of them, I imagine, were amazed to see the first batch of 400 sell within the first hour and a half. Despite how intensely annoying it is, the single-dollar toy went on to sell 300 million units, making James a very wealthy man. In 1960 he packed everything up, left his wife to run the business and moved to Bolivia, where he joined the cult of the Wycliffe Bible Translators. And that's where he stayed until his death in 1974.

THE MILLION-DOLLAR HOMEPAGE

In 2005 a twenty-one-year-old British student, Alex Tew, came up with a novel way of raising the money for his university education. He set up a single-page website that offered to display one million adverts (one per pixel) for a single dollar each. Amazingly, www.milliondollarhome page.com was an instant success and advertisers poured in to buy pixels on the webpage. Alex shamelessly declared himself a 'pixel hustler and proud of it', and then he sat back to watch the money roll in. Launched on 25 August 2005, the final pixels were auctioned on 11 January 2006 and the sales income achieved was $1,037,100 against start-up costs of under $75 to register the domain name. It was a stroke of genius: stand up and take a bow, young man.

Atomic energy might be as good as our present day explosives, but it is unlikely to produce anything very much more dangerous.
Sir Winston Churchill, British Prime Minister,
in 1939

EPONYMOUS INVENTIONS

MAXIM GUN

Hiram Stevens Maxim (1840–1916) was an American inventor who began his working life as a coachbuilder. He later worked as a draughtsman and instrument maker, and he also took a keen interest in his brother Hudson's career as a military inventor specializing in explosives. In 1881 Maxim emigrated to England where he established himself in London and became a British subject. At that time he was working as an electrical engineer (he claimed to have invented the light bulb in 1878) and the following year had an epiphany that would change his life. And end thousands of others. He later recalled: 'In 1882 I was in Vienna where I met an American who I had previously known. He said to me, "Hang your electricity and chemistry. If you want to make piles of money then invent something that will help these Europeans cut each other's throats with greater efficiency."'

Maxim pondered the advice and began to take a closer look at the rifles of the day, noting that the powerful recoil and time they took to reload were obvious limitations. One day, whilst out on the rifle range, Maxim realized that if he could use the energy created by recoil to automatically reload the weapon he could create the ultimate rapid fire and self-loading rifle. He was now in the machine-gun business. In June 1883 Maxim registered his patents, and by October of the following year he was ready to demonstrate his new weapon to military leaders around the world. Generals and politicians were invited to fire the Maxim Gun themselves, and a clever marketing campaign described the weapon as a 'great peace-maker'. One famous claim by Maxim himself was that his gun could 'fell trees'. Demonstrations proved that it could.

The British government ordered three but, although the design passed all of the stipulated tests, the military high command predicted limited use for the gun. So Maxim turned to the emerging European power, the newly formed Germany, and arranged a demonstration that was observed by Kaiser Wilhelm II himself. The Kaiser was impressed and authorized its use. The Maxim Gun was distributed throughout the German Army as the *Maschinengewehr*, although it first saw action with the British, who had ordered a few more and used four of them to repel 4,000 African warriors during the First Matabele War for Rhodesia in 1893–4. Despite positive reports from the battlefield the British remained unconvinced, and even as late as 1915, as the First World War was warming up, General Douglas Haig announced, 'The machine gun is a much overrated weapon. Two per battalion is more than sufficient.' It was eventually redesigned, improved and used by both sides as the Great War claimed millions of lives. Far from being a peace-maker, the Maxim Gun was probably the most lethal invention of the nineteenth century.

THE LEOTARD
Where would gymnasts and ballerinas be today if it wasn't for the great French trapeze artist Jules Léotard

(1842–70)? Famous for his performances with the Cirque Franconi in Paris (later the Cirque Napoleon), Léotard was also the inspiration for the popular song 'The Darling Young Man on the Flying Trapeze' by George Leybourne that was popularized in 1867. Léotard was born in Toulouse, the son of a gymnastics instructor, but he showed little interest in the family business as a youngster, preferring to study law instead. Eventually, at the age of eighteen, he began practising with ropes, rings and bars over a swimming pool, and soon developed into one of the most popular entertainers of his generation. However, he will be remembered, and some say honoured, for all time for his invention of the one-piece signature costume that he had knitted together to allow both freedom of movement and little chance of loose material tangling in ropes or rings. Performers quickly copied the costume that, to this day, bears the name of the great Jules Léotard.

THE JACUZZI

Anybody who has ever said, 'I don't need a Jacuzzi, I just eat beans before taking a bath,' must stand up and leave the room immediately. This is a serious business. The Jacuzzi brothers (there were seven of them) emigrated to the United States, from Italy, around the turn of the twentieth century. They all trained as machinists and in 1915, after being inspired by the Panama Pacific Exposition of San Francisco, the eldest brother, Rachele Jacuzzi, decided to start working on designs for aircraft propellers. His wooden Jacuzzi Toothpick propeller was an immediate success and the brothers decided to open a manufacturing company in Berkeley called Jacuzzi Brothers.

One of their early designs was a closed cabin mono-plane that was soon used by the US Postal Service all over America, but the Jacuzzis soon lost their appetite for aviation when, in 1921, one of their planes crashed, killing brother Giocondo. Instead, the remaining six turned their attention to hydraulic pumps that had been developed for the aviation industry but which Rachele recognized could be utilized throughout both industry and agriculture. Their revolutionary design for deep-well agricultural pumps won them the Gold Medal at the California State Fair in 1930.

> Flight by machines that are heavier than air is unpractical and insignificant, if not utterly impossible.
> Simon Newcomb, Canadian-born American astronomer, in 1902

By 1948 one of the brothers, Candido, used the company technology to design an underwater pump to provide home hydrotherapy treatment for his son Kenneth, who was suffering rheumatoid arthritis. He had been receiving the treatment at a local hospital and Candido hated to see the boy suffering between sessions. Jacuzzi Brothers began marketing the J-300 pump in 1955 as a therapeutic aid and were soon advertising it as a remedy for 'worn out housewives'. Film stars Jane Mansfield and Randolph Scott endorsed the Jacuzzi Pump and by 1958 the firm were selling self-contained Jacuzzi bathtubs, which were soon popular throughout Hollywood. Hundreds of thousands were sold as the Jacuzzi quickly became the luxury item of choice throughout America. Kenneth Jacuzzi grew up to take over the company from his uncles, and under his steward-ship the firm grew into an international concern that was bought, in October 2006, for just under $1 billion. Today the Jacuzzi Group Worldwide is still a major supplier of spas and hot-tubs all over the world.

THE GUILLOTINE

It was the French Revolution that made a certain Dr Joseph-Ignace Guillotin (1738–1814) world famous, although ironically he did not invent the machine that caused nearly 40,000 French heads to roll. Originally known as a *louisette* after the name of its real creator, the guillotine was actually invented by a military surgeon called Antoine Louis; Guillotin himself was publicly opposed to capital punishment.

In 1784, when Franz Mesmer published his theory of animal magnetism, the French public were so offended by his conclusions that Louis XVI established a commission, which included the eminent physician Dr Guillotin, to investigate the matter. In 1789, following publication the previous year of his pamphlet *Petition of the Citizens Living in Paris*, Guillotin was appointed one of the ten Paris Deputies of the Assemblée Constituante. During a debate on capital punishment, he proposed that 'the criminal should be decapitated and solely by means of a simple mechanism, a machine that beheads painlessly'. It was Guillotin's belief that, if an execution had to be carried out, it should at least be pain free. His initiative led to proposals for a new 'death machine', and the development of the *louisette*, later renamed the 'guillotine'.

There is a common misconception that Guillotin

was himself eventually executed by the machine that bore his name. Although he was arrested and imprisoned for a short time, he was released in 1794. The Dr J. M. V. Guillotin who was dispatched at Lyon during the early 1790s was a different person altogether. Our Dr Guillotin retired from politics and returned to the safety of the medical profession, becoming one of the founders of the Academy of Medicine in Paris. Guillotin was also one of the most vocal supporters of Edward Jenner's theory of vaccination, which has saved millions of lives over the last two centuries.

After Guillotin's death from natural causes in 1814, his family, embarrassed by the infamous death machine, petitioned the French government to change its name. When their appeal was turned down, they changed their name instead, and have lived in peaceful obscurity ever since. The guillotine, on the other hand, remained in service until September 1977, when a one-legged Tunisian became its final victim.

> Computers in the future may weigh no more than 1.5 tons.
>
> *Popular Mechanics*, predicting the progress of technology in 1949

THE DAVY LAMP

The eldest of five children, Humphry Davy was born in Penzance, Cornwall, on 17 December 1778. A brilliant scientist, he became a fellow of the Royal Society and professor of the Royal Institution. Davy was hugely popular with the public and his lecture tours and experiments were always well attended, although in 1812 laboratory accidents cost him two fingers and the use of one eye.

In 1814 Sir Humphry Davy (he had been knighted two years previously) settled back into his laboratory and, with the Felling Mine Disaster of 1812 at a colliery near Newcastle in mind, began work to improve the underground pit lighting and safety of miners. By 1815 he had produced a safety lamp that enabled miners to work deep seams despite the presence of methane or other flammable gases. At the time, all mines were illuminated by the naked flame, and explosions were a constant hazard. But Davy discovered that a flame enclosed by a fine wire mesh could not ignite any dangerous gases (known as 'firedamp') as air could pass through the gauze, keeping the flame alight, but the holes in the mesh were too fine to allow the flame to pass the other way and ignite the firedamp. In addition, the flame inside the safety lamp would burn with a blue tinge if any firedamp was present. Placed near the ground, the lamp could also be used to detect denser gases, such as carbon monoxide, the invisible killer. If there was insufficient oxygen in the air, the flame would go out, acting as an early warning for miners to evacuate. No doubt canary lovers the world over were just as pleased as the miners by Davy's innovative ideas. (A canary in a cage had previously provided the early warning system – a dead bird indicating the presence of toxic gases.)

Despite his many other significant scientific discoveries, it is the miner's safety lamp, and his contribution to the welfare of the men who worked the mines, for which Davy is best remembered, and to this day pubs in former mining communities all over Britain still bear the name of his life-saving invention.

SHRAPNEL: MAXIMUM DAMAGE

Shrapnel is a fascinating word that sounds as though it must have been around forever – one of those words in English clearly reflecting its Scandinavian roots. Sadly, this impression is completely wrong, for the word itself is far more recent in origin.

It has evolved in meaning too. The modern dictionary definition of 'shrapnel' is 'fragments of an exploding bomb', whereas originally, during the First World War, it meant the whole explosive device, not just parts of it. 'Shrapnel-shells' were designed as anti-personnel artillery, packed with bullets that would discharge close to the target with the obvious intention of killing or maiming as many of the enemy as possible. Shrapnel-shells were far more effective for the purpose than a conventional bomb, but they became obsolete at the end of the war when they were replaced with high-explosive shells, which did much the same job and whose deadly fragments were still known as 'shrapnel'.

The original First World War weapon was named after Major General Henry Shrapnel (1761–1842), a British Army officer and inventor who served in the Royal Artillery during the Napoleonic Wars (1803–15). He devised a hollow cannonball filled with grapeshot, which was attached to a rocket and designed to burst in midair, creating multiple casualties. Shrapnel's idea was not to kill enemy soldiers but to maim them, since a dead man needs no immediate attention whereas a wounded one requires the attention of at least two others, if only to remove him from the battlefield. Such was the success of his invention that Henry Shrapnel was awarded over £1,000 in 1814, a considerable sum of money in those days, and in 1827 was promoted to the post of Colonel Commandant of the Royal Artillery.

The first verse of the American national anthem proudly describes the Americans resisting an onslaught of shrapnel by the British Army during the pivotal Battle of Baltimore in 1812: 'And the rockets' red glare, the bombs bursting in air, Gave proof through the night that our flag was still there.'

Men might as well project a voyage to the moon as attempt to employ steam navigation against the stormy North Atlantic Ocean.

Dr Dionysus Lardner (1793–1859), Professor of Natural Philosophy and Astronomy at University College London

---

THE DIESEL ENGINE

Rudolf Diesel (1858–1913) was born in Paris, the son of a German bookbinder. At the outbreak of the Franco-Prussian War (1870), the family were forced to leave their home, along with most other German natives in France, and fled to London instead of returning east. However, they did send twelve-year-old Rudolf back to their home town of Augsburg to live with his aunt and uncle, the mathematics professor Christoph Barnickel. After graduating top of his class, Rudolf enrolled in the Royal Bavarian Polytechnic of Munich, against the wishes of his parents who wanted him to return to London and find a job to help support the family. Instead, Rudolf studied under the German engineer Carl von Linde, the pioneer of refrigeration, although he failed to graduate after falling ill with typhoid and missing his examination. But he persevered, using his time to study practical engineering, and finally graduated in 1880 at the age of twenty-two, before joining von Linde, who was by then himself in Paris building a refrigeration and ice plant. Within one year Diesel was offered the position of managing director, and one of his first decisions was to develop a more efficient engine and power supply than the steam engines the industry relied upon at the time.

The main problem with steam engines was that their loss of heat and energy meant they only provided about 10 per cent of the power of which they were capable. Diesel set about creating a new engine that would transfer as much of an engine's energy as possible into useful work, and so he started experimenting with current engines in an effort to discover a way to modify them. Von Linde supported his research and the company filed many patents during the course of Diesel's work. However, early attempts proved to be disastrous, and almost fatal, as one of his test engines exploded and almost killed him. After many months in hospital, Rudolf Diesel returned to his work with a new idea. He had remembered from his younger days how his bicycle pump heated up at the valve when compressed air, from the piston mechanism, was forced into the tyre.

By 1891 von Linde had lost patience with his protégé and the two men parted company. Diesel was forced to find new funding to continue his work and between 1893 and 1897 Heinrich von Buz, an engineer from the Diesel family's home town of Augsburg, provided the facilities. Finally, in 1895, Rudolf Diesel was granted a patent in Germany and the United States for his compression-ignition piston engine that worked, largely, along the lines of a bicycle pump, with a piston that would force air to become hot enough to ignite fuel, which would propel the piston back down to repeat the cycle.

It was a huge advancement on the steam engines that had been used for the previous 200 years. More importantly, he was just in time to influence the growing car and forthcoming aviation industries. Neither of which would have been possible without Diesel's perseverance and bravery, not to mention his bicycle pump. And he knew it too, writing to his wife: 'I am now so far ahead of everything that has been achieved that, in the manufacture of engines on our little planet, I now lead the field on both sides of the ocean.'

With his patents secured, the thirty-seven-year-old inventor became a very wealthy man as his engines were soon being built all over the industrial world. But his wealth was to be short lived as expensive legal battles in defending his patent, poor investments and his family's lavish lifestyle all began to take their toll. The money was running out quickly and, when he became aware of it, Diesel arranged a series of crisis meetings at his manufacturing headquarters in London and with his financiers.

On the evening of 29 September 1913, Diesel was on his way to London on board the mail steamer SS *Dresden* when he retired to his cabin, after dining, at 10 p.m., asking to be called at 6.15 a.m. The following morning crew members reported to the captain that there was no sign of the famous inventor. His cabin was empty, his bed had not been slept in and his nightshirt was neatly laid upon it. His pocket watch was on the bedside, his hat and overcoat tidily stored. Rudolf Diesel was never seen again. A Dutch steamer retrieved a corpse from the North Sea ten days later but it was too badly decomposed to identify.

The crew removed all personal belongings and buried the body at sea. A few days later Rudolf's son, Eugen, identified the items as belonging to his father. Suicide emerged as the most likely explanation, especially as the only entry he made in his diary for that day was a black cross. He had also given his wife a bag with instructions not to open it until the following week. Inside were a number of bank statements, all with balances of virtually zero, and 200,000 German marks in cash. However, murder was never ruled out as his business and military interests may have provided a motive. A mysterious end for a man whose engines were, by that time, powering manufacturing plants, locomotive engines, cars, lorries, airships, aeroplanes, submarines and ships. Today, a century later, Rudolf Diesel's engine remains one of the most important sources of power on the planet.